Saas-Fee Advanced Course 38

More information about this series at http://www.springer.com/series/4284

T. L. Wilson · Stéphane Guilloteau

Millimeter Astronomy

Saas-Fee Advanced Course 38

Swiss Society for Astrophysics and Astronomy
Edited by Miroslava Dessauges-Zavadsky
and Daniel Pfenniger

Authors
T. L. Wilson
Max-Planck-Institut für Radioastronomie
Bonn
Germany

Stéphane Guilloteau
Laboratoire d'Astrophysique de Bordeaux
Université Bordeaux 1
Bordeaux
France

Volume Editors
Miroslava Dessauges-Zavadsky
Observatoire de Genève
Université de Genève
Sauverny
Switzerland

Daniel Pfenniger
Observatoire de Genève
Université de Genève
Sauverny
Switzerland

This Series is edited on behalf of the Swiss Society for Astrophysics and Astronomy: Société Suisse d'Astrophysique et d'Astronomie, Observatoire de Genève, ch. des Maillettes 51, CH-1290 Sauverny, Switzerland.

Cover illustration: A montage of ALMA Antennae, Credit ESO, with Barnard 150 Dark Nebula in Cepheus, http://www.astrogb.com/b150.htm, © Giovanni Benintende, Reproduced with permission

ISSN 1861-7980 ISSN 1861-8227 (electronic)
Saas-Fee Advanced Course
ISBN 978-3-662-58602-0 ISBN 978-3-662-57546-8 (eBook)
https://doi.org/10.1007/978-3-662-57546-8

© Springer-Verlag GmbH Germany, part of Springer Nature 2018
Softcover re-print of the Hardcover 1st edition 2018
This work is subject to copyright. All rights are reserved by the Publisher, whether the whole or part of the material is concerned, specifically the rights of translation, reprinting, reuse of illustrations, recitation, broadcasting, reproduction on microfilms or in any other physical way, and transmission or information storage and retrieval, electronic adaptation, computer software, or by similar or dissimilar methodology now known or hereafter developed.
The use of general descriptive names, registered names, trademarks, service marks, etc. in this publication does not imply, even in the absence of a specific statement, that such names are exempt from the relevant protective laws and regulations and therefore free for general use.
The publisher, the authors and the editors are safe to assume that the advice and information in this book are believed to be true and accurate at the date of publication. Neither the publisher nor the authors or the editors give a warranty, express or implied, with respect to the material contained herein or for any errors or omissions that may have been made. The publisher remains neutral with regard to jurisdictional claims in published maps and institutional affiliations.

Printed on acid-free paper

This Springer imprint is published by the registered company Springer-Verlag GmbH, DE
part of Springer Nature
The registered company address is: Heidelberger Platz 3, 14197 Berlin, Germany

Preface

The millimeter (mm) and sub-millimeter (sub-mm) wavebands are unique in astronomy in containing more than 1000 spectral lines of molecules as well as the thermal continuum spectrum of cold dust. They are the only bands in the electromagnetic spectrum in which we can detect cold dust and molecules far away in high-redshift galaxies, and nearby in low-temperature cocoons of protostars and protoplanets.

Observations in mm/sub-mm wavebands promise to make a decisive contribution to the following key questions in the current astronomy:

- The origins of galaxies: Current optical studies are limited to the very brightest objects. In mm and sub-mm wavebands, it is possible to detect galaxies 100 times fainter and dust-obscured out to the epoch of reionization. The redshifts of these galaxies can be measured directly and precisely, either photometrically, based on the shape of the spectral energy distribution, or spectroscopically, using the many available spectral lines. A complete picture of the star formation history of the universe requires the knowledge of discrete sources that produce the far-infrared/sub-mm background.
- The evolution of galaxies: Images of the molecular gas in galaxies at the resolution of the Hubble Space Telescope (HST) will give the information on both the parsec and kiloparsec scales needed to explore the relationship between star formation, gas density, and gas kinematics, in comparison with other tracers, like the atomic gas, Hα, or radio continuum. The role of density waves and spiral structure, the mechanisms of starbursts and the associated feedback processes (such as outflows of molecular gas, bubbles, and winds), and the effects of mergers can hence be addressed.
- Star and planet formation: The mm/sub-mm wavebands are ideal for studying how gas and dust evolve from a collapsing molecular cloud core into a circumstellar disk that can form planets by providing unique information on the kinematics and mass distribution inside the cores and their envelopes. When the newly formed stars are surrounded by protoplanetary disks, imaging the gas and

dust on scales of several astronomical units is the only way to study the earliest stages of planet formation.

Given these important prospects and with the largest mm/sub-mm facility, the Atacama Large Millimeter/Sub-millimeter Array (ALMA), having started its operation, the members of the Swiss Society for Astrophysics and Astronomy chose this topic for the 38th Saas-Fee advanced course. ALMA is undoubtedly producing a major step in astrophysics comparable to that provided by the HST and will work in synergy with the James Webb Space Telescope.

The course took place in the small Swiss village of Les Diablerets, Switzerland. The selected lecturers, T. L. Wilson, Stéphane Guilloteau, and Pierre Cox, have offered to about 60 participants outstanding and pedagogical lectures on the millimeter observational techniques and the above scientific topics. We wish to sincerely thank them for their successful and high scientific level course. We express special thanks to T. L. Wilson and Stéphane Guilloteau whose determination and hard work in writing their respective chapters enabled to assemble this book.

Finally, we warmly thank the course secretary, Myriam Burgener Frick, for all her help, enthusiasm, and devotion in the practical organization.

Geneva, Switzerland
2017

Miroslava Dessauges-Zavadsky
Daniel Pfenniger

Contents

Introduction to Millimeter/ Sub-millimeter Astronomy

T. L. Wilson

Contents

T. L. Wilson (✉)
Max Planck Institute for Radio Astronomy, Auf dem Hügel 69,
53121 Bonn, Germany
e-mail: twilson@mpifr-bonn.mpg.de

© Springer-Verlag GmbH Germany, part of Springer Nature 2018

M. Dessauges-Zavadsky and D. Pfenniger (eds.), *Millimeter Astronomy*,
Saas-Fee Advanced Course 38, https://doi.org/10.1007/978-3-662-57546-8_1

1 Introduction

This chapter provides an introduction to the basics of radiative transfer, receivers, antennas and interferometry. Following this is an exposition of radiation mechanisms for continuum, atoms and molecules. Much of this material is contained in [36, 52]. This chapter places more emphasis on current research in millimeter/sub-millimeter astronomy. The field of mm/sub-mm astronomy has become very large, so this presentation contains only a few derivations. In some cases, the approach is to quote a result followed by an example. Some common terms are used, but "jargon" has been avoided as much as possible. For the most part, references are to more recent work, where citations to earlier publications can be found. The units are mostly CGS with some SI units. This follows the usage in the astronomy literature. One topic *not* covered here is the polarization (see [36, 47] for an introduction). Another glaring omission is a treatment of the Cosmic Microwave Background (CMB), since this is not treated in the following lecture. The CMB emission is not weak, but does fill the entire sky, so special techniques and different interpretations must be employed.

The field of mm/sub-mm astronomy began very recently, but has produced a richness of results. Millimeter/sub-mm measurements require excellent weather, very accurate antennas, and sensitive receivers, so there are relatively few such facilities. The interpretation of these data requires a knowledge of atomic and molecular physics, radiative transfer and interstellar chemistry.

A fundamental discovery of mm/sub-mm astronomy was the recognition that molecular clouds exist. By number, about 90% of molecular clouds consist of H_2 (with about 10% in helium). However, the determination of local densities and column densities of molecular hydrogen, H_2 must be indirect since H_2 does not emit spectral lines in cooler clouds. The interpretation of molecular line astronomy data often requires an understanding of interstellar chemistry. The Schrödinger equation governs all chemistry, but the conditions in the interstellar medium are very different (and much more varied) than those on earth. Earth-bound chemistry is only a subset of the more general interstellar chemistry; it is dominated by non-equilibrium processes that occur at low temperatures and densities, so determinations of collision and reaction rates are needed. For an overview of interstellar chemistry, the reader is referred to the presentation [18] or the monograph [45]. Although interstellar chemistry plays a very important role in molecular line astronomy, this is *not* included due to space limitations.

Synchrotron emission is more intense at meter or centimeter wavelengths with a rapid fall off in intensity at mm/sub-mm wavelengths. Measurements of synchrotron emission from Active Galactic Nuclei (AGN's) allows sub-arcsecond resolution images of relativistic electrons moving in $\boldsymbol{B}$ fields [41]. The intensities of dust and thermal emission from molecules are at most the kinetic temperature, so are small. These measurements are best carried out in the mm/sub-mm range where intensities are larger. Thus the imaging of thermal emission from dust or spectral lines on arcsecond scales requires high sensitivity. This is a major reason for the construction of the *Atacama Large Millimeter Array* (ALMA). With ALMA, images of thermal emission on arcsecond (or better) scales is an area where important breakthroughs will be made.

The dominant broadband continuum radiation mechanism in the mm/sub-mm wavelength range is dust emission. This differs from free-free and synchrotron emission, which is commonly encountered at cm wavelengths [26]. Spectral line radiation is dominated by thermal and quasi-thermal molecular radiation. In addition, there are a few important atomic lines of carbon, oxygen and nitrogen.

In the field of star formation, it is believed that all of the physical principles are known. However, star formation is complex and measurements are needed to determine which processes are dominant and which can be neglected. High resolution imaging of gas and dust is one of the most important tools to study the birth of stars and planets [35, 44]. Stars form in molecular clouds where extinction is large, so near infrared and optical studies are of limited value.

The study of nearby star forming regions is complemented with studies of star formation in more distant galaxies [43]. At higher redshifts, star forming galaxies and Active Galactic Nuclei (AGN's) are enshrouded by dust [42]. The images of such regions in spectral lines and dust continuum in the mm/sub-mm range will complement the images already made in the cm wavelength range. These results may have a great impact on our interpretation of such regions. In contrast to star formation, for studies of relativistic phenomena such as the the early universe, Black Holes and AGN's, not all of the relevant physical laws may be known. As in all of astronomy, measurements often lead to unexpected results that may have far reaching effects on our understanding of fundamental physical laws. The most interesting example of relativistic physics in the mm/sub-mm range is Sgr A*. This is thought to be the closest supermassive Black Hole, at 8.5 kpc (1 kpc = 3.08×10^{21} cm) from the sun [32]. The radio emission from Sgr A* has been interpreted as optically thick synchrotron emission, which becomes optically thin at about 0.8 mm. Sgr A* is thought to be similar to the "engines" that power AGN's. With Very Long Baseline Interferometry (VLBI) in the mm/sub-mm range, one can sample radiation close to the Schwarzschild radius.

Turning to technical matters, the receiver sensitivity at mm/sub-mm wavelengths is more than 100 times better than in the 1960's [17]. However sensitivity alone is not enough to transform the field. Rather, high resolution images are needed. This will change when the *Atacama Large Millimeter Array* begins full operation. ALMA will initially operate with receiver bands in all atmospheric windows between 10 and 0.4 mm. With a unique combination of high angular resolution and high sensitivity, ALMA will transform astronomy. A summary of the science planned for ALMA is to be found in [4].

2 Some Background

In this section, we review the basics needed in following sections (see [36]). Electromagnetic radiation in the radio window can be interpreted as a wave phenomenon, i.e., in term of classical physics. When the scale of the system involved is much larger than a wavelength, we can consider the radiation to travel in straight lines or *rays*. The power, $\mathrm{d}P$, radiated from an object of angular size $\mathrm{d}\Omega$, which is intercepted by an infinitesimal surface $\mathrm{d}\sigma$ is

$$\mathrm{d}P = I_\nu \cos\theta \,\mathrm{d}\Omega \,\mathrm{d}\sigma \,\mathrm{d}\nu \,, \tag{1}$$

where

$\mathrm{d}P$ = power, in Watts,
$\mathrm{d}\sigma$ = area of surface, m^2,
$\mathrm{d}\nu$ = bandwidth, in Hz,
θ = angle between the normal to $\mathrm{d}\sigma$ and the direction to $\mathrm{d}\Omega$,
I_ν = brightness or specific intensity, in $\mathrm{W\,m^{-2}\,Hz^{-1}\,sr^{-1}}$.

Equation (1) is the definition of the brightness I_ν. Quite often the term *intensity* or *specific intensity* I_ν is used instead of the term *brightness*. We will use all three designations interchangeably.

The total flux density of a source is obtained by integrating Eq. (1) over the total solid angle Ω_s subtended by the source

$$S_\nu = \int_{\Omega_s} I_\nu(\theta, \varphi) \cos\theta \, d\Omega \ , \tag{2}$$

where the source solid angle is characterized by the usual angles θ and φ. This flux density is measured in units of W m^{-2} Hz^{-1}. Since the flux density of astronomical sources is usually very small, a special unit, the Jansky (hereafter Jy) has been introduced

$$1 \text{ Jy} = 10^{-26} \text{ W m}^{-2} \text{ Hz}^{-1} = 10^{-23} \text{ erg s}^{-1} \text{ cm}^{-2} \text{ Hz}^{-1} \ . \tag{3}$$

As long as the surface element $d\sigma$ covers the ray bundle completely, the power remains constant:

$$dP_1 = dP_2 \ . \tag{4}$$

From this, we can easily obtain

$$I_{\nu_1} = I_{\nu_2} \tag{5}$$

so that the brightness is independent of the distance. The total flux S_ν density shows the expected dependence of $1/r^2$ (where r is the radius).

Another useful quantity related to the brightness is the radiation energy density u_ν in units of erg cm^{-3}. From dimensional analysis, u_ν is intensity divided by speed. Since radiation propagates with the velocity of light c, we have for the *spectral energy density per solid angle*

$$u_\nu(\Omega) = \frac{1}{c} I_\nu \ . \tag{6}$$

If integrated over the whole sphere, 4π steradian, Eq. (6) will yield the *total spectral energy density*

$$u_\nu = \int_{(4\pi)} u_\nu(\Omega) \, d\Omega = \frac{1}{c} \int_{(4\pi)} I_\nu \, d\Omega \ . \tag{7}$$

2.1 Radiative Transfer

The *equation of transfer* is

$$\boxed{\frac{dI_\nu}{ds} = -\kappa_\nu I_\nu + \varepsilon_\nu} \ . \tag{8}$$

where ds is the distance. The linear absorption coefficient κ_ν and the emissivity ε_ν are independent of the intensity I_ν leading to the above form for dI_ν.

In Thermodynamic Equilibrium (TE) there is complete equilibrium of the radiation with its surroundings, the brightness distribution $B_\nu(T)$ is described by the Planck function, which depends only on the temperature T of the surroundings. The properties of the Planck function will be described in the next section.

In Local Thermodynamic Equilibrium (LTE), the *Kirchhoff's law* holds

$$\boxed{\frac{\varepsilon_\nu}{\kappa_\nu} = B_\nu(T)} \; . \tag{9}$$

This is independent of the material, as is the case with complete thermodynamic equilibrium. In general however, I_ν will differ from $B_\nu(T)$.

If we define the *optical depth* $\mathrm{d}\tau_\nu$ by

$$\mathrm{d}\tau_\nu = -\kappa_\nu \, \mathrm{d}s \tag{10}$$

or

$$\tau_\nu(s) = \int_{s_0}^{s} \kappa_\nu(s)\, \mathrm{d}s \;, \tag{11}$$

then the equation of transfer (8) can be written as

$$\boxed{-\frac{1}{\kappa_\nu}\frac{\mathrm{d}I_\nu}{\mathrm{d}s} = \frac{\mathrm{d}I_\nu}{\mathrm{d}\tau_\nu} = I_\nu - B_\nu(T)} \; . \tag{12}$$

The solution of Eq. (12) is

$$\boxed{I_\nu(s) = I_\nu(0)\,\mathrm{e}^{-\tau_\nu(s)} + \int_{0}^{\tau_\nu(s)} B_\nu(T(\tau))\,\mathrm{e}^{-\tau}\mathrm{d}\tau} \; . \tag{13}$$

If the medium is isothermal,

$$T(\tau) = T(s) = T = \text{const.}$$

the integral is

$$\boxed{I_\nu(s) = I_\nu(0)\,\mathrm{e}^{-\tau_\nu(s)} + B_\nu(T)\,(1 - \mathrm{e}^{-\tau_\nu(s)})} \; . \tag{14}$$

For a large optical depth, that is for $\tau_\nu(0) \to \infty$, Eq. (14) in LTE approaches the limit

$$I_\nu = B_\nu(T) \; . \tag{15}$$

This is case for planets and the 2.7 K microwave background. The difference between $I_\nu(s)$ and $I_\nu(0)$ gives

$$\Delta I_\nu(s) = I_\nu(s) - I_\nu(0) = (B_\nu(T) - I_\nu(0))(1 - e^{-\tau}) . \qquad (16)$$

Equation (16) represents an on-source minus an off-source measurement.

2.2 *Black Body Radiation and Brightness Temperature*

The spectral distribution of the radiation of a black body in thermodynamic equilibrium is given by the Planck law

$$\boxed{B_\nu(T) = \frac{2h\nu^3}{c^2} \frac{1}{e^{h\nu/kT} - 1}} .$$

If $h\nu \ll kT$, one obtains *Rayleigh -Jeans Law*

$$\boxed{B_{\rm RJ}(\nu, T) = \frac{2\nu^2}{c^2} kT} . \qquad (17)$$

This is an expression of classical physics in which there is an energy kT per mode. The term $\frac{2\nu^2}{c^2}$ is the *density of states* for 3 dimensions. In the millimeter and sub-millimeter range, one frequently defines a *radiation temperature* $J(T)$ as

$$J(T) = \frac{c^2}{2k\nu^2} I = \frac{h\nu}{k} \frac{1}{e^{h\nu/kT} - 1} . \qquad (18)$$

Inserting numerical values for k and h, we find that the Rayleigh–Jeans relation holds for frequencies

$$\frac{\nu}{\rm GHz} \ll 20 \left(\frac{T}{\rm K}\right) . \qquad (19)$$

It can thus be used for all thermal radio sources except perhaps for low temperatures in the mm/sub-mm range.

In the Rayleigh–Jeans relation, the brightness and the thermodynamic temperature of the black body that emits this radiation are strictly proportional (Eq. (17)). This feature is so useful that it has become the custom in radio astronomy to measure the brightness of an extended source by its *brightness temperature* $T_{\rm b}$. This is the temperature which would result in the given brightness if inserted into the Rayleigh–Jeans law

$$T_{\rm b} = \frac{c^2}{2k} \frac{1}{\nu^2} I_\nu = \frac{\lambda^2}{2k} I_\nu \ . \tag{20}$$

Combining Eq. (2) with Eq. (20), we have

$$\boxed{S_\nu = \frac{2\,k\,\nu^2}{c^2} T_{\rm b}\,\Delta\Omega} \ . \tag{21}$$

For a Gaussian source, this relation is

$$\left[\frac{S_\nu}{\rm Jy}\right] = 0.0736\,T_{\rm b} \left[\frac{\theta}{\text{arc seconds}}\right]^2 \left[\frac{\lambda}{\rm mm}\right]^{-2} \ . \tag{22}$$

That is, if the flux density S_ν and the source size are known, then the true brightness temperature $T_{\rm b}$ of the source can be determined. The concept of temperature in radio astronomy has given rise to confusion. If one measures S_ν and the *apparent* source size, Eq. (22) allows one to calculate the *main beam brightness temperature* $T_{\rm MB}$. Details of this procedure will be given in Sect. 5. The performance of coherent receivers is characterized by the *receiver noise temperature*; see Sect. 3.2.1 for details. The combination of receiver and atmosphere is characterized by the *system noise temperature*; this is discussed in Sect. 6.1 and following.

If I_ν is emitted by a black body and $h\nu \ll kT$ then Eq. (20) gives the thermodynamic temperature of the source, a value that is independent of ν. If other processes are responsible for the emission of the radiation, $T_{\rm b}$ will depend on the frequency; it is, however, still a useful quantity and is commonly used in practical work.

This is the case even if the frequency is so high that condition (19) is not valid. Then Eq. (20) can still be applied, but $T_{\rm b}$ is different from the thermodynamic temperature of a black body. However, it is rather simple to obtain the appropriate correction factors.

It is also convenient to introduce the concept of brightness temperature into the radiative transfer Eq. (14). Formally one can obtain

$$J(T) = \frac{c^2}{2k\nu^2}(B_\nu(T) - I_\nu(0))(1 - {\rm e}^{-\tau_\nu(s)}) \ .$$

Usually calibration procedures allow one to express $J(T)$ as T. This measured quantity is referred to as $T_{\rm R}^*$, the *radiation temperature*, or the *brightness temperature*, $T_{\rm b}$. In the cm wavelength range, one can apply Eqs. (20)–(12) and one obtains

$$\boxed{\frac{{\rm d}T_{\rm b}(s)}{{\rm d}\tau_\nu} = T_{\rm b}(s) - T(s)} \ , \tag{23}$$

where $T(s)$ is the thermodynamic temperature of the medium at the position s. The general solution is

$$\boxed{T_{\mathrm{b}}(s) = T_{\mathrm{b}}(0)\,\mathrm{e}^{-\tau_\nu(s)} + \int\limits_0^{\tau_\nu(s)} T(s)\,\mathrm{e}^{-\tau}\mathrm{d}\tau} \ . \tag{24}$$

If the medium is isothermal, this becomes

$$\boxed{T_{\mathrm{b}}(s) = T_{\mathrm{b}}(0)\,\mathrm{e}^{-\tau_\nu(s)} + T\,(1 - \mathrm{e}^{-\tau_\nu(s)})} \ . \tag{25}$$

2.3 The Nyquist Theorem and the Noise Temperature

We now relate voltage and temperature; this is essential for the analysis of receiver systems limited by noise. The average power per unit bandwidth produced by a resistor R is

$$P_\nu = \langle i\,v \rangle = \frac{\langle v^2 \rangle}{2R} = \frac{1}{4R}\langle v_{\mathrm{N}}^2 \rangle \ , \tag{26}$$

where $v(t)$ is the voltage that is produced by i across R, and $\langle \cdots \rangle$ indicates a time average. The first factor $\frac{1}{2}$ arises from the condition for the transfer of maximum power from R over a broad range of frequencies. The second factor $\frac{1}{2}$ arises from the time average of v^2. An analysis of the random walk process shows that

$$\langle v_{\mathrm{N}}^2 \rangle = 4R\,k\,T \ . \tag{27}$$

Inserting this into Eq. (26) we obtain

$$P_\nu = k\,T \ . \tag{28}$$

Equation (28) can also be obtained by a formulation of the Planck law for one dimension and the Rayleigh–Jeans limit. Then, the available noise power of a resistor is proportional to its temperature, the *noise temperature* T_{N}, and independent of the value of R. Throughout the whole radio range, from the longest waves to the far infrared region the noise spectrum is white, that is, its power is independent of frequency. Since the impedance of a noise source should be matched to that of the receiver, such a noise source can only be matched over some finite bandwidth.

2.3.1 Hertz Dipole and Larmor Formula

In contrast to thermal noise, radiation from an antenna has a definite polarization and direction. One example (in electromagnetism there are *no* simple examples!) is a Hertz dipole of length l. In the far field, the total power radiated from a Hertz dipole carrying an oscillating current I at a wavelength λ is

$$\boxed{P = \frac{2c}{3}\left(\frac{I\,\Delta l}{2\lambda}\right)^2}\ . \tag{29}$$

For the Hertz dipole, the radiation is linearly polarized with the $\boldsymbol{E}$ field along the direction of the dipole. The power pattern has a donut shape, with a cylindrically symmetric maximum perpendicular to the axis of the dipole. Along the direction of the dipole, the radiation field is zero. One can use collections of dipoles, driven in phase, to restrict the direction of radiation. As a rule of thumb, Hertz dipole radiators have the best radiative efficiency when the wavelength of the radiation is roughly the size of the dipole. Following Planck, one can produce the the Black Body law from a collection of (quantized) Hertz dipole oscillators with random phases.

There are similarities between Hertz dipole radiation and radiation from atoms. Following Larmor, the power radiated by a single electron oscillating with a velocity $v(t)$ is

$$P(t) = \frac{2}{3}\frac{e^2\dot{v}(t)^2}{c^3}\ . \tag{30}$$

Expressing $\dot{v} = \ddot{x}$ and $x = \sin 2\pi\nu t$, we obtain an average power:

$$\langle P\rangle = \frac{64\pi^4}{3\,c^3}\,\nu_{mn}^4\left(\frac{e\,x_0}{2}\right)^2\ . \tag{31}$$

Using $\frac{e\,x_0}{2} = |\mu|$ one arrives at the expression for spontaneous emission from a quantum mechanical system. This is presented again in Eq. (132). It is often the case that quantum mechanical expressions and classical correspond to within factors of a few.

3 Signal Processing and Stationary Stochastic Processes

Next, we present the basics of signal processing and noise analysis needed to understand the properties of radiometers [36]. Fourier methods are of great value in analyzing receiver properties [6]. A review of sub-millimeter receiver systems is to be found in [37].

The concept of spectral power density was introduced in Eq. (26) in a practical example. Radio receivers are devices that measure spectral power density. Since the signals are dominated by noise, statistical analyses are needed. The most important of these statistical quantities is the probability density function, $p(x)$, which gives the probability that at any arbitrary moment of time the value of the process $x(t)$ falls within an interval $(x - \frac{1}{2}\,\mathrm{d}x, x + \frac{1}{2}\,\mathrm{d}x)$. For a stationary random process, $p(x)$ will be independent of the time t.

The *expected value* $E\{x\}$ or *mean value* of the random variable x is given by the integral

$$\boxed{E\{x\} = \int_{-\infty}^{\infty} x\, p(x)\,\mathrm{d}x} \tag{32}$$

and, by analogy, the expectation value $E\{f(x)\}$ of a function $f(x)$ is given by

$$\boxed{E\{f(x)\} = \int_{-\infty}^{\infty} f(x)\, p(x)\,\mathrm{d}x}\,. \tag{33}$$

The trend in *all* forms of communications (including radio astronomy!) is toward digital processing. In general, a digitized function must be *sampled* at regular intervals. Assume that the input signal extends from zero Hz to ν_0 Hz (this is referred to the video band). Then the bandwidth, $\Delta\nu_0$, extends from 0 Hz to ν_0, the maximum frequency. If we picture the input as a collection of sine waves, it is clear that the sampling rate *must* be at least $\nu_0 = 2\Delta\nu$ to characterize the sinusoid with the highest frequency, $\sin 2\pi\nu_0 t$. Thus the sample rate must be twice the bandwidth of the video input. This is referred to as the *Nyquist Sampling Rate*. This is a minimum; a higher sampling rate can only improve the characterization of the input. A higher sampling rate ("oversampling") will allow the input to be better characterized, thus giving a better Signal-to-Noise ratio (S/N) ratio. The sampling functions must occupy an extremely small time interval compared to the time between samples. If only a portion of the input function is retained in the quantization and sampling process, information is lost. This results in a lowering of the S/N ratio. At present, commercially available digitizers can sample a 1.5 GHz bandwidth with 8 bit quantization. This allows digital systems to reach within a few percent of the sensitivity of analogue systems, but with much greater stability. After digitization, one can maintain that all of the following processes are "just arithmetic". The quality of the data (i.e., the S/N) depends on the analogue receiver elements. The digital parts of a receiver can lower S/N ratios, but not raise them!

3.1 Square Law Detectors

In radio receivers, noise with an RMS standard deviation σ is passed through a device that produces an output signal $y(t)$ which is proportional to the power in a given input

$v(t)$:

$$y(t) = a\, v^2(t)\ . \tag{34}$$

The assumption is that $v(t)$ follows a Gaussian distribution, with an Expectation Value, E. The calculation of E involves an evaluation of the integral

$$E\{y(t)\} = E\{a\, v^2(t)\} = \frac{a}{\sigma\sqrt{2\pi}} \int_{-\infty}^{+\infty} v^2 \mathrm{e}^{-v^2/2\sigma^2} \mathrm{d}v\ .$$

There are standard ways used to evaluate this expression. The result is

$$E\{y(t)\} = E\{a\, v^2(t)\} = a\, \sigma_v^2\ . \tag{35}$$

For the evaluation of $E\{y^2(t)\}$, one must calculate

$$E\{y^2(t)\} = E\{a^2\, v^4(t)\} = \frac{a^2}{\sigma\sqrt{2\pi}} \int_{-\infty}^{+\infty} v^4 \mathrm{e}^{-v^4/2\sigma^2} \mathrm{d}v\ .$$

The result of this integration is

$$E\{y^2(t)\} = 3\, a^2\, \sigma_v^4 \tag{36}$$

and hence

$$\sigma_y^2 = E\{y^2(t)\} - E^2\{y(t)\} = 2\, E^2\{y(t)\}\ . \tag{37}$$

3.2 Limiting Receiver Sensitivity

A receiver must be *sensitive*, that is, be able to detect faint signals in the presence of noise. There are limits for this sensitivity, since the receiver input and the receiver itself are affected by noise. Even when no input source is connected to a receiver, there is an output signal, since any receiver generates thermal noise. This noise is amplified together with the signal. Since signal and noise have the same statistical properties, these cannot be distinguished. To analyze the performance of a receiver we will use the model of an ideal receiver producing no internal noise, but connected simultaneously to two noise sources, one for the external source noise and a second for the receiver noise. To be useful,receivers must increase the input power level. The power per unit bandwidth, P_ν, entering a receiver can be characterized by a temperature as given by Eq. (28), $P_\nu = kT$. For the receiver itself, this is T_R. Furthermore, it is *always* the case that the noise contributions from source, atmosphere, ground and receiver, T_i, are additive,

$$T_\mathrm{sys} = \sum T_i\ .$$

An often used figure of merit is the *Noise Factor*, F. This is defined as

$$F = \frac{S_1/N_1}{S_2/N_2} = \frac{N_2}{G\,N_1} = 1 + \frac{T_R}{T_1} \tag{38}$$

that is, any additional noise generated in the receiver contributes to N_2. For direct detection systems, such as a *Bolometers*, $G = 1$. If T_1 is set equal to $T_0 = 290$ K, we have

$$T_R = (F - 1) \cdot 290 \,.$$

Given a value of F, one can determine the receiver noise temperature. If for $\nu_0 =$ 115 GHz, $F = 3$ db ($= 10^{0.1\,F\ \mathrm{db}}$), $T_R = 290$ K, a lousy receiver noise temperature.

3.2.1 Receiver Calibration

Our goal is to characterize receiver noise performance in degrees Kelvin. In the calibration process, a noise power scale (spectral power density) is established at the receiver input. In radio astronomy the noise power of coherent receivers (i.e., those which preserve the phase of the input) is usually measured in terms of the noise temperature. In this process, the receiver is assumed to be a linear power measuring device (i.e., we assume that any non-linearity of the receiver is a small quantity). To calibrate a receiver, one relates the noise temperature increment ΔT at the receiver input to a given measured receiver output increment Δz (this applies to coherent receivers which have a wide dynamic range and a total power or "DC" response). In principle, the receiver noise temperature, T_R, could be computed from the output signal z provided the detector characteristics are known. In practice, the receiver is calibrated by connecting two or more known sources to the input. Usually matched resistive loads at known (thermodynamic) temperatures T_L and T_H are used. To within a constant, the receiver outputs are

$$z_L = (T_L + T_R)\,G\,,$$
$$z_H = (T_H + T_R)\,G\,,$$

from which

$$\boxed{T_{rx} = \frac{T_H - T_L\,y}{y - 1}}\,, \tag{39}$$

where

$$y = z_H/z_L \,. \tag{40}$$

This is known as the "y-factor"; the procedure is a "hot-cold" measurement. Note that the y-factor as presented here is determined in the Rayleigh–Jeans limit. The temperatures T_H and T_L are usually produced by absorbers in the mm/sub-mm range. Usually these are chosen to be at the ambient temperature ($T_H \cong 293$ K or 20 °C) and at the temperature of liquid nitrogen ($T_L \cong 78$ K or −195 °C). In rare cases, one might use liquid helium, which has a boiling point $T_L \cong 4.2$ K. Usually such a "hot-cold" calibration is done infrequently. As will be discussed in Sect. 6.2 in the mm/sub-mm wavelength range, measurements of the emission from the atmosphere and then from an ambient resistive load are combined with models to provide an estimate of the atmospheric transmission. For a determination of the receiver noise, an additional measurement, usually with a cooled resistive load is needed.

Bolometers (Sect. 4.1) do not preserve phase, so are incoherent receivers. Their performance is strongly dependent on the bias of the detector element. The Bolometer performance is characterized by the *Noise Equivalent Power*, or NEP. The NEP is given in units of Watts $\mathrm{Hz}^{-1/2}$. NEP is the input power level which doubles the output power. Usually bolometers are "AC" coupled, that is, the output responds to *differences* in the input power, so hot-cold measurements are not useful for characterizing bolometers (see Fig. 2 for the schematic of a bolometer).

3.2.2 Noise Uncertainties Due to Random Processes

It has been found that both source and receiver noise has a Gaussian distribution [36]. We assume that the signal is a Gaussian random variable with mean zero

$$p(x) = \frac{1}{\sigma\sqrt{2\pi}}\, \mathrm{e}^{-x^2/2\sigma^2} \tag{41}$$

which is sampled at a rate equal to twice the bandwidth.

Figure 1 shows a receiver block diagram. By assumption $E(v_1) = 0$. The input, v_1 has a much larger bandwidth, B, than the bandwidth of the receiver, that is, $\Delta\nu \ll B$. The output of the receiver is v_2, with a bandwidth $\Delta\nu$. The power corresponding to the voltage v_2 is $\langle v_2^2 \rangle$.

$$P_2 = v_2^2 = \sigma^2 = k\, T_{\mathrm{sys}}\, G\Delta\nu\ , \tag{42}$$

where $\Delta\nu$ is the receiver bandwidth, G is the gain, and T_{sys} is the total noise from the input T_A and the receiver T_R. The contributions to T_A are the external inputs from the source, ground and atmosphere. Given that the output of the square law detector is v_3

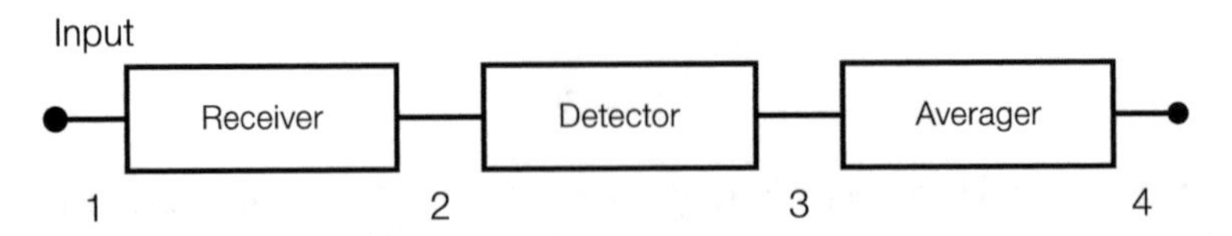

Fig. 1 The principal parts of a receiver. In the text, v_1 is at point 1, v_2 at point 2, etc

$$\langle v_3 \rangle = \left\langle v_2^2 \right\rangle \tag{43}$$

then after square-law detection we have

$$\langle v_3 \rangle = \left\langle v_2^2 \right\rangle = \sigma^2 = kT_{\rm sys} G \Delta\nu \,. \tag{44}$$

Crucial to a determination of the noise is the mean value and variance of $\langle v_3 \rangle$. From Eq. (36) the result is

$$\left\langle v_3^2 \right\rangle = \left\langle v_2^4 \right\rangle = 3 \left\langle v_2^2 \right\rangle \,, \tag{45}$$

this is needed to determine $\left\langle \sigma_3^2 \right\rangle$. Then,

$$\sigma_3^2 = \left\langle v_3^2 \right\rangle - \langle v_3 \rangle^2 \tag{46}$$

$\left\langle v_3^2 \right\rangle$ is the total noise power (= receiver plus input signal). Using the Nyquist sampling rate, the averaged output, v_4, is $(1/N)\Sigma v_3$ where $N = 2\Delta\nu\,\tau$.

From v_4 and $\sigma_4^2 = \sigma_3^2/N$, we obtain the result

$$\sigma_4 = k\Delta\nu G (T_{\rm A} + T_{\rm R})/\sqrt{\Delta\nu\,\tau} \,. \tag{47}$$

We have explicitly separated $T_{\rm sys}$ into the sum $T_{\rm A} + T_{\rm R}$. Finally, we make use of the calibration procedure in Sect. 3.2.1 to express the term $kG\Delta\nu$ as a temperature. Then

$$\boxed{\frac{\Delta T}{T_{\rm sys}} = \frac{1}{\sqrt{\Delta\nu\,\tau}}} \,. \tag{48}$$

The calibration process allows us to specify the receiver output in degrees Kelvin instead of in Watts per Hz. We therefore characterize the receiver quality by the system noise temperature $T_{\rm sys} = T_{\rm A} + T_{\rm R}$. τ corresponds to the exposure time.

For a given system, the improvement in the RMS noise *cannot* be better than as given in Eq. (48). Systematic errors will only increase ΔT, although the time behavior may follow the behavior described by Eq. (48) [9]. Thus $T_{\rm sys}$ is the noise from the *entire* system. That is, it includes the noise from the receiver, atmosphere, ground, and *the source*. At some wavelengths, the earth's atmosphere may contribute significant noise. For measurements of some planets in the mm/sub-mm range, ΔT is larger for an intense source.

3.2.3 Receiver Stability

Sensitive receivers are designed to achieve a low value for $T_{\rm sys}$. Since the signals received are of exceedingly low power, receivers must also provide sufficient output power. This requires a large receiver gain, so even very small gain instabilities can dominate the thermal receiver noise. Therefore, receiver stability considerations are also of prime importance. Great advances have been made in improving receiver stability. However, in the mm/sub-mm range, the atmosphere plays an important role. To insure that the noise decreases following Eq. (48), systematic effects from atmospheric and receiver instabilities are minimized. Atmospheric *changes* are of crucial importance. These can be compensated for by rapidly taking the difference between the measurement of the source of interest and a reference region or a nearby calibration source. Such *comparison switched* measurements are necessary for all ground based observations. R. H. Dicke was the first to apply comparison switching to radio astronomical receivers [9].

The time spent measuring references or performing calibrations will *not* contribute to an improvement in the S/N ratio. In fact, the subtraction of two noisy inputs *worsens* the difference, but are needed to reduce instabilities that give rise to systematic errors. The time τ is the *total* time taken for the measurement (i.e., on-source and off-source).

Even for the output of a total power receiver there will be additional noise in excess of that given by Eq. (48) since the signals to be differenced are $\Delta T + T_{\rm sys}$ and $T_{\rm sys}$. This is needed since $\Delta T << T_{\rm sys}$. For example, if one-half the total time is spent on the reference, the ΔT for difference of on-source minus off-source in Eq. (48) is a factor of 2 larger.

4 Practical Receivers

This section is concerned with the practical aspects of receivers that are currently in use [13, 37], with some background material from [36].

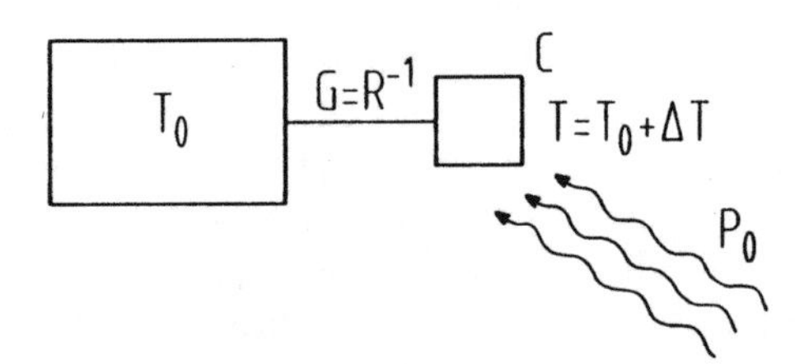

Fig. 2 A bolometer is represented by the smaller square to the right. The power from an astronomical source, P_0, raises the temperature of the bolometer element by ΔT, which is much smaller than the temperature T_0 of the heat sink. Heat capacity, $\mathcal{C}$, is analogous to capacitance. The conductance, $\mathcal{G}$, is analogous to electrical conductance, G, which is $1/R$. The noise performance of bolometers depends critically on the thermodynamic temperature, T_0, and on the conductance $\mathcal{G}$. The temperature change causes a change in the voltage drop across the bolometer (the details of the electric circuit are not shown)

4.1 Bolometer Radiometers

The operation of bolometers makes use of the effect that the resistance, R, of a material varies with the temperature (see Fig. 2). When radiation is absorbed by the bolometer material, the temperature varies; this temperature change is a measure of the intensity of the incident radiation. Because this thermal effect is rather independent of the frequency of the radiation absorbed, bolometers are intrinsically broadband devices. Frequency discrimination must be provided by external filters.

The Noise Equivalent Power *(NEP)* quoted for a bolometer is the input power that doubles the output of this device. The expression for NEP is

$$\boxed{\mathrm{NEP_{ph}} = 2\varepsilon\, k\, T_{\mathrm{BG}}\, \sqrt{\Delta\nu}}\,, \tag{49}$$

where ε is the emissivity of the background, and T_{BG} is temperature of the background. Typical values for ground based bolometers are $\varepsilon = 0.5$, $T_{\mathrm{BG}} = 300$ K and $\Delta\nu = 100$ GHz. For these values $\mathrm{NEP_{ph}} = 1.3 \times 10^{-15}$ Watts $\mathrm{Hz}^{-1/2}$. With the collecting area of the 30-m IRAM telescope and a 100 GHz bandwidth one can easily detect mJy sources.

4.2 Currently Used Bolometer Systems

Bolometers mounted on ground based radio telescopes are background noise limited, so the only way to substantially increase mapping speed for extended sources is to construct large arrays consisting of many pixels. In present systems, the pixels are separated by 2 beamwidths, because of the size of individual bolometer feeds. The systems which best cancel atmospheric fluctuations are composed of rings of close-packed detectors surrounding a single detector placed in the center of the array. Two large bolometer arrays have produced many significant published results. The first is MAMBO2 (MAx-Planck-Millimeter Bolometer). This is a 117 element array used at the IRAM 30-m telescope at Pico Veleta, Spain. This system operates at 1.3 mm, and provides an angular resolution of 11″. The portion of the sky that is measured at one instant is the *field of view* (FOV). The FOV of MAMBO2 is 240″. The second system is SCUBA (Sub-millimeter Common User Bolometer Array) [21]. This is used on the James-Clerk-Maxwell (JCMT) 15-m sub-mm telescope at Mauna Kea, Hawaii. SCUBA consists of a 37 element array operating at 0.87 mm with an angular resolution of 14″ and a 91 element array operating at 0.45 mm with an angular resolution of 7.5″; both have a FOV of about 2.3′. The LABOCA (LArge Bolometer CAmera) array operates on the APEX 12-m telescope. APEX is on the 5.1 km high Chaijnantor plateau, the ALMA site in northern Chile. The LABOCA camera operates at 0.87 mm wavelength with 295 bolometer elements. These are arranged in 9 concentric hexagons around a center element. The angular resolution of each element is 18.6″, the FOV is 11.4′. Such an arrangement is ideal for the measurement of small sources since the outer rings of detectors can be used to subtract the emission from the sky, while the central elements are measuring the source.

4.2.1 Superconducting Bolometers

A promising new development in bolometer receivers is *Transition Edge Sensors* referred to as TES bolometers. These superconducting devices may allow more than an order of magnitude increase in sensitivity, if the bolometer is not background limited. For broadband bolometers used on earth-bound telescopes, the warm background limits the performance. With no background, the noise improvement with TES systems is limited by photon noise; in a background noise limited situation, TES's should be $\sim 2-3$ times more sensitive than semiconductor bolometers. For ground based telescopes, TES's greatest advantage is multiplexing many detectors with a superconducting readout device, so one can construct even larger arrays of bolometers. SCUBA has been replaced with SCUBA-2 that has been constructed at the U. K. Astronomy Technology Center. SCUBA-2 is an array of 2 TES bolometers, each consisting of 6400 elements operating at 0.87 and 0.45 mm. The FOV of SCUBA-2 is 8′. The SCUBA-2 design is based on photo-deposition technology similar to that used for integrated circuits. This type of construction allows for a closer packing of the individual bolometer pixels. In SCUBA-2 these are separated by 1/2 of a beam, instead of the usual 2 beam spacing. Additional such systems are SHARC on the Caltech 10-meter telescope on Mauna Kea, Hawaii, and the MUSTANG array on the 100-meter Green Bank Telescope.

4.2.2 Polarization and Spectral Line Measurements

In addition to measuring the continuum total power, one can mount a polarization-sensitive device in front of the bolometer and thereby measure the direction and degree of linear polarization. These devices have been used with SCUBA on the James-Clerk-Maxwell Telescope in Hawaii.

It is possible to also carry out spectroscopy, if frequency sensitive elements, either Michelson or Fabry-Perot interferometers, are placed before the bolometer element. Since these spectrometers operate at the sky frequency, the fractional resolution $(\Delta\nu/\nu)$ is limited.

4.3 Coherent Receivers

Coherent receivers are those which preserve the phase of the signal. Usually, coherent receivers make use of the superheterodyne (or more commonly "heterodyne") princi-

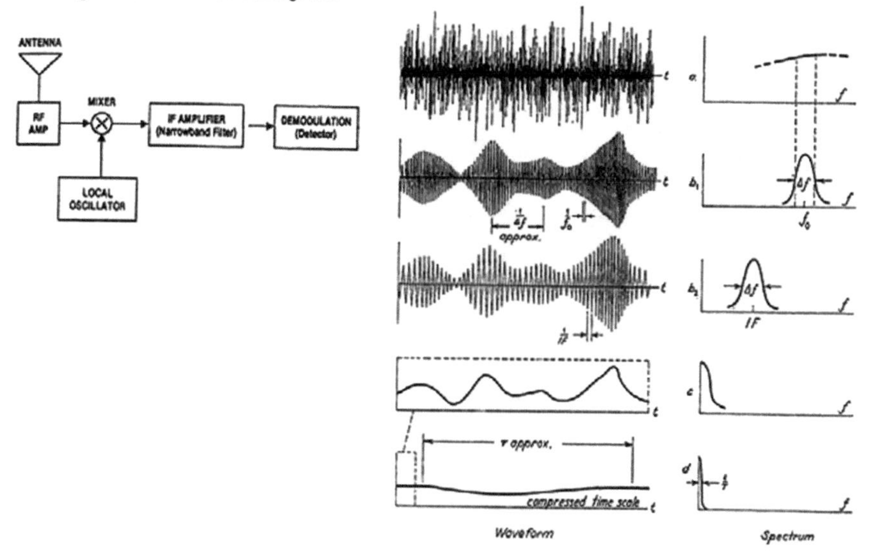

Fig. 3 On the left are the elements of a coherent receiver. For $\nu_0 < 120$ GHz one can amplify before mixing while for higher frequencies the first circuit element is the mixer. In the middle are sketches of the time behavior of broadband noise after passing through these parts of the receiver; on the right are sketches of the corresponding frequency behavior. The mixer shifts the signal to a lower frequency while the amplifier increases the output over a narrow band. The square law detector converts rapidly oscillating signals into a smooth response that has positive values. The middle and right sketches are taken from Pawsey and Bracewell (1954) [34]

ple to shift the signal input frequency without changing other properties; in practice, this is carried out by the use of mixers (Sect. 4.3.3). Heterodyne is commonly used in all branches of communications technology; its use allows measurements with unlimited spectral resolution, $\Delta\nu/\nu$. A schematic of a heterodyne receiver is shown in Fig. 3.

4.3.1 The Minimum Noise in a Coherent System

The ultimate limit for coherent receivers or amplifiers is obtained by application of the *Heisenberg uncertainty principle*. The *minimum* noise of a coherent amplifier results in a receiver noise temperature of

$$\boxed{T_{\rm rx}(\text{minimum}) = \frac{h\nu}{k}} \,. \tag{50}$$

For *incoherent* detectors, such as bolometers, phase is not preserved, so this limit does *not* exist. In the millimeter wavelength regions, this noise temperature limit is quite small. At $\lambda = 2.6\,\mathrm{mm}$ ($\nu = 115\,\mathrm{GHz}$), this limit is 5.5 K.

4.3.2 Elements of Coherent Receivers

The noise in the first element dominates the system noise. The exact expression is given by the *Friis* relation which takes into account the effect of having cascaded amplifiers:

$$\boxed{T_\mathrm{S} = T_\mathrm{S1} + \frac{1}{G_1} T_\mathrm{S2} + \frac{1}{G_1 G_2} T_\mathrm{S3} + \cdots + \frac{1}{G_1 G_2 \ldots G_{n-1}} T_{\mathrm{S}n}}\,, \tag{51}$$

where G_1 is the gain of the first element, and T_S1 is the noise temperature of this element. The corresponding values apply to the following elements in a receiver. For lossy elements, G is less than unity. For $\lambda < 3$ mm, cooled first elements typically have $G_1 = 10^3$ and $T_\mathrm{S1} = 50$ K; for $\lambda < 0.3$ mm, cooled first elements typically have $G_1 = 1$ and $T_\mathrm{S1} = 500$ K.

4.3.3 Mixers

Mixers allow the signal frequency to be changed without altering the characteristics of the signal. In the mixing process, one multiplies the input signal with an intense monochromatic signal from a *local oscillator*, LO, in a non-linear circuit element. In principle a mixer produces a shift in frequency of an input signal with no other effect on the signal properties. For a single mixer, two frequency bands, at equal

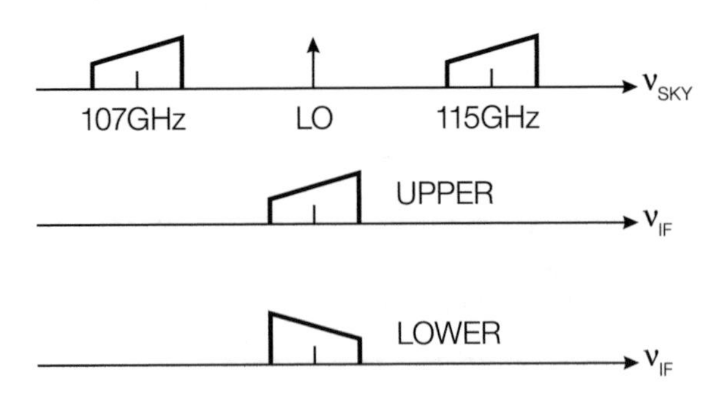

Fig. 4 A sketch of the frequencies shifted from the sky frequency (top) to the output (lower) of a double sideband mixer. In this example, the input is at the sky frequencies for the Upper Side Band (USB) of 115 GHz, and Lower Side Band (LSB) of 107 GHz while the output frequency is 4 GHz. The slanted boxes represent the passbands; the direction of the slant in the boxes indicate the upper (higher) and lower (lower) edge of the bandpass in frequency. ALMA mm mixers are SSB mixers so these effects are avoided. Figure taken from Wilson et al. (2013) [52]

separations from the LO frequency are shifted into intermediate (IF) frequency band. This is Double Sideband (DSB) mixer operation (see Fig. 4). These are referred to as the *signal* and *image* bands. In the mm/sub-mm wavelength ranges, such mixers are still commonly used as the first stage of a receiver. For single dish *continuum* measurements, both sidebands contain the signal, so DSB operation does not decrease the signal-to-noise (S/N) ratio. However, for single dish spectral line measurements, the spectral line of interest is in one sideband only. The other sideband is then a source of extra noise (i.e., lower S/N ratio) and perhaps confusing lines. Therefore, single sideband (SSB) operation is desired. If the image sideband is eliminated, the mixer is said to operate in SSB mode. This can be accomplished by inserting a filter before the mixer. However, filters cause a loss of signal, so lower the S/N ratio. For low noise applications a more complex arrangement is needed. In many cases, a *single sideband* mixer is used. This consists of two identical mixers driven by a single local oscillator through a phase shifting device.

For example, the ALMA Band 7 mm/sub-mm receiver shifts a signal centered at 345 GHz from the sky frequency, 341–349 GHz to 4 – 12 GHz. At 4 to 12 GHz the signal is easily amplified. This is an SSB mixer; the unwanted sideband is not accepted.

Noise in mixers has 3 causes. The first is the mixer itself. Since one half of the input signal at $\nu_{\rm sky}$ is shifted to an unwanted frequency $\nu_{\rm LO} + \nu_{\rm sky}$, the input signal is lowered by a factor-of-two (3 db). This is referred to as *conversion loss*. Classical mixers typically have 3 db (=factor of 2) loss. In Eq. (51), $G_1 = 1/2$. In addition there will be an additional noise contribution from the mixer itself ($T_{\rm S1}$ in Eq. (51)). Second, the LO may have "phase noise", that is a rapid change of phase, which will affect signal properties, so adds to the uncertainties. Third, the amplitude of the LO may vary. This effect can be minimized since the mixer LO power is adjusted so that the mixer output is saturated. Then there is no variation of the output signal power if LO power varies.

4.4 HEMTs/MMICs

Within a highly ordered crystal made of identical atoms, free electrons can move only within certain *energy bands*. By varying the material, both the width of the band, the *band gap*, and the energy to reach a conduction band can be varied. The crucial part of any semiconductor device is the junction. On the one side there is an excess of material with negative carriers, forming n-type material and the other side material with a deficit of electrons, that is p-type material. The p-type material has an excess of positive carriers, or *holes*. At the junction of a p- and n-type material, the electrons in the n-type material can diffuse into the p-type material (and vice-versa), so there will be a net potential difference. The diffusion of charges, p to n and n to p, cannot continue indefinitely, but a difference in the charges near the boundary of the n and p material will remain, because of the low conductivity of the semiconductor material. From the potential difference at the junction, a flow of electrons in the positive direction is easy, but a flow in the negative direction will be hindered. Typical p-n junctions have a slow response so are suitable only as square-law detectors. Schottky (metal-semiconductor) junctions have a lower capacitance, so are better suited to applications such as microwave mixers. The combination of three layers, p-n-p, in a so-called "sandwich", is a simple extension of the p-n junction. In Field Effect Transistors, FET's, the electric field controls the carrier flow so this is an amplifier. At frequencies of 100 GHz, unipolar devices, which have only one type of carrier, are used as microwave amplifier front ends. *High Electron Mobility Transistors*, HEMTs, are an evolution of FETs. The design goals of HEMT's are: (1) to obtain lower intrinsic amplifier noise and (2) operation at higher frequency. In HEMTs, the charge carriers are present in a channel of small size. This confinement gives rise to a two dimensional electron gas, or "2 DEG", where there is less scattering and hence lower noise. When cooled, there is a significant improvement in the noise performance, since the main contribution is from the oscillations of nuclei in the lattice, which are strongly temperature dependent. To extend the operation of HEMT to higher frequencies, one must increase electron mobility, μ, and saturation velocity V_s. A reduction in the scattering by introducing impurities ("doping") leads to a larger electron mobility, μ, and hence faster transit times, in addition to lower amplifier noise.

For use up to $\nu = 115$ GHz with good noise performance, one has turned to modifications of HEMTs based on advances in material-growth technology. The SEQUOIA receiver array of the Five College Radio Astronomy Observatory uses Microwave Monolithic Integrated Circuits (MMIC's) in 32 front ends for a 16 beam, two polarization system (pioneered by S. Weinreb). The MMIC is a complete amplifier on a single semiconductor, instead of using lumped components. The MMIC's have excellent performance in the 80–115 GHz region without requiring tuning adjustments. The simplicity makes MMIC's better suited for multi-beam systems.

For low noise IF amplifiers, 4–8 GHz IF systems using Gallium-Arsinide HEMTs with 5 K noise temperature and more than 20 db of gain have been built. With Indium-Phosphide HEMTs on GaAs-substrates, even lower noise temperatures are possible. As a rule of thumb, one expects an increase of 0.7 K per GHz for GaAs, while the corresponding value for InP HEMTs is 0.25 K per GHz. For front ends, noise temperatures of the amplifiers in the 18–26 GHz range are typically 12 K.

4.4.1 Superconducting Mixers

Very general, semi-classical considerations indicate that the slope of the current-voltage, $I - V$, curve for classical mixers changes gently. This leads to a relatively poor noise figure for classical mixers, since much of the input signal is not converted to a lower frequency.

A significant improvement can be obtained if the junction is operated in the superconducting mode. Then the gap between filled and empty states is comparable to photon energies in the mm/sub-mm range. In addition, the LO power requirements are $\approx$1000 times lower than are needed for conventional mixers. Finally, the physical layout of such devices is simpler since the mixer is a planar device, deposited on a substrate by lithographic techniques. SIS mixers consist of a superconducting layer, a thin insulating layer and another superconducting layer. SIS mixers depend on single carriers; a longer but more accurate description of SIS mixers is "single quasiparticle photon assisted tunnelling detectors". When the SIS junction is properly biased, the filled states reach the level of the unfilled band, and the electrons can quantum mechanically tunnel through the insulating strip. The $I - V$ curve for a SIS device shows sudden jumps in the $I - V$ curve; these are typical of quantum-mechanical phenomena. For low noise operation, the SIS mixer must be DC biased at an appropriate voltage and current. If, in addition to the mixer bias, there is a source of photons of energy $h\nu$, then the tunnelling can occur. If one then biases an SIS device and applies an LO signal at a frequency ν, the $I - V$ curve becomes very sharp, so the conversion of sky signals to the IF frequency is much more effective than with a classical mixer.

Under certain circumstances, SIS devices can produce gain. If the SIS mixer is tuned to produce substantial gain, it is unstable. Thus, this is not useful in radio astronomical applications. In the mixer mode, that is, as a frequency converter, SIS devices can have a small amount of gain. This tends to balance inevitable mixer losses, so SIS devices have losses that are lower than classical mixers. SIS mixers have performance that is unmatched in the mm/sub-mm region. In addition to *single sideband* properties, improvements to existing designs include *tunerless* and SIS mixers. Tunerless mixers have the advantage of repeatability in tuning. For ALMA, SIS mm mixer designs are wideband, tunerless, single sideband devices with extremely low mixer noise temperatures.

An increase in the gap energy is needed to allow the efficient mixing at higher frequencies. This is done with Niobium superconducting materials; the geometric junction sizes are 1 μm by 1 μm. For frequencies above 900 GHz, one uses niobium nitride junctions. Variants of such devices, such as the use of junctions in series, can be used to reduce the capacitance. An alternative is to reduce the size of the individual junctions to 0.25 μm.

SIS mixers are the front ends of choice for operation between 150 GHz and 900 GHz because these are low-noise devices, the IF bandwidths can be $>$1 GHz, these are tunable over $\sim$30% of the frequency range and the local oscillator power needed is $<$1 μW.

4.4.2 Hot Electron Bolometers

Superconducting Hot Electron Bolometer-mixers (HEB) are heterodyne devices, in spite of the name. These mixers make use of superconducting thin films which have sub-micron sizes. In an HEB mixer excess noise is removed either by diffusion of hot electrons out the junction, or by an electron-phonon exchange. The first HEBs operating on radio telescopes and used to take astronomical made use of electron-phonon exchange. The HEB junctions were of μm size, consisting of Niobium Nitride (NbN), cooled to 4.2 K. Junctions of Aluminum-Titanium-Nitride, AlTiN, have provided lower receiver noise temperatures. For these devices, the IF center frequency was 1.8 GHz and had a full width of 1 GHz. Such a system was used to measure the $J = 9 - 8$ carbon monoxide line at 1.037 THz.

4.4.3 Single Pixel Receiver Systems

In summary, devices that provide the lowest noise front ends are:

- for $\nu < 115$ GHz, High Electron Mobility Transistors (HEMT) and Microwave Monolithic Integrated Circuits (MMIC)
- for $72 < \nu < 800$ GHz, Superconducting Mixers (SIS)
- for $\nu > 900$ GHz, Hot Electron Bolometers (HEB)

SIS mixers provide the lowest receiver noise in the mm and sub-mm range. SIS mixers are much more sensitive than classical Schottky mixers, and require less local oscillator power, but must be cooled to 4 K. All millimeter mixer receivers are tunable over 10–20% of the sky frequency. From the band gaps of junction materials, there is a short wavelength limit to the operation of SIS devices. For spectral line measurements at wavelengths $\lambda < 0.3$ mm, superconducting Hot Electron Bolometers, which have no such limit, have been developed. At frequencies above 2 THz there is a transition to far-infrared and optical techniques. The highest frequency heterodyne systems in radio astronomy are used in the Herschel-HIFI satellite. These are SIS and HEB mixers.

The SIS or HEB mixers convert the sky frequency to the fixed IF frequency, where the signal is amplified by the IF amplifiers. Most of the amplification is done in the IF. The IF should only contribute a negligible part to the system noise temperature. Because some losses are associated with frequency conversion, the first mixer is a major source for the system noise. Two ways exist to decrease this contribution: (1) by use of either an SIS or HEB mixer to convert the input to a lower frequency, or (2) at lower frequencies by use of a low-noise amplifier before the mixer.

4.4.4 Multibeam Systems

At 3 mm, the SEQUOIA array receiver produced at the Five College Radio Astronomy Observatory (FCRAO) with 32 MMIC front ends connected to 16 beams had been used on 14-m telescope of the FCRAO for the last few years. Multibeam system that use SIS front ends are rare. A 9 beam Heterodyne Receiver Array of SIS mixers at 1.3 mm, HERA, has been installed on the IRAM 30-m millimeter telescope to measure spectral line emission. To simplify data taking and reduction, the HERA beams are kept on a Right Ascension-Declination coordinate frame. HARP-B is a 16 beam SIS system in operation at the James-Clerk-Maxwell telescope. The sky frequency is 325–375 GHz. The beam size of each element is 14″, with a beam separation of 30″, and a FOV of about 2′. The total number of spectral channels in a heterodyne multi-beam system will be large. In addition, complex optics is needed to properly illuminate all of the beams. In the mm range this usually means that the receiver noise temperature of each element is somewhat larger than that for a single pixel receiver system. For further details of SEQUOIA, HERA, or HARP, see the appropriate web sites.

For single dish continuum measurements at $\lambda < 2$ mm, multi-beam systems make use of bolometers. GeGa bolometers are the most common systems and the best such systems have a large number of beams. In the future, TES bolometers have great promise. Compared to incoherent receivers, heterodyne systems are still the most efficient for spectral lines in the range $\lambda > 0.3$ mm, although Fabry-Perot systems (such as SPIFI; see the web site) may be competitive for some projects. For bolometers on the Herschel satellite, one uses gratings or Fabry-Perot systems. For SCUBA-2, an analog Michelson (Fourier transform interferometer) is proposed.

4.5 Back Ends: Spectrometers

The term "Back End" is used to specify the devices following the IF amplifiers. Of the many different back ends that have been designed for specialized purposes, spectrometers are probably the most widely used. Previously this was carried out in especially designed hardware, but recently there have been devices based on general purpose digital computers.

Spectrometers analyze the spectral information contained in the radiation field. To accomplish this these must be SSB and the frequency resolution $\Delta\nu$ is usually fine; perhaps in the kHz range. In addition, the time stability must be high. If a resolution of $\Delta\nu$ is to be achieved for the spectrometer, all those parts of the system that enter critically into the frequency response have to be maintained to better than $0.1\,\Delta\nu$. An overview of the current state of spectrometers is to be found in [5].

4.5.1 Multichannel Filter Spectrometers

The time needed to measure the power spectrum for a given celestial position can be reduced by a factor n if the IF section consists of n contiguous filters covering the bandwidth $\Delta\nu$. After each filter there is a square-law detector and the integrator. These must be built not merely once, but n times. Then these form n separate channels that simultaneously measure different (usually adjacent) parts of the spectrum. Such a system is rather cost intensive so alternatives have been sought.

4.6 Fourier, Autocorrelation and Cross Correlation Spectrometers

One method is to Fourier Transform (FT) the input, $v(t)$, to obtain $v(\nu)$ and then square $v(\nu)$ to obtain the Power Spectral Density. From the Nyquist theorem, it is necessary to sample at a rate equal to twice the bandwidth. These are referred to as "FX" autocorrelators. Recent developments at the Jodrell Bank Observatory have led to the building of COBRA (Coherent Baseband Receiver for Astronomy). COBRA can analyze a 100 MHz bandwidth. A similar device with a 1 GHz bandwidth has been built at the Max-Planck-Institut in Bonn for use in the mm/sub-mm range on the APEX telescope.

For Autocorrelators, or XF systems, the input $v(t)$ is correlated with a delayed signal $v(t-\tau)$ to obtain the autocorrelation function $R(\tau)$. $R(\tau)$ is then Fourier Transformed to obtain the spectrum. For an XF system the time delays are performed in a set of serial digital shift registers with a sample delayed by a time τ. Autocorrelation can also be carried out with the help of analog devices using a series of cable delay lines. Such analogue correlators have been developed at the University of Maryland together with NRAO for use on the Green Bank Telescope (GBT); these are used to provide very large bandwidths.

The two significant advantages of digital spectrometers are: (1) flexibility and (2) a noise behavior that follows $1/\sqrt{t}$ after many hours of integration. The flexibility allows one to select many different frequency resolutions and bandwidths or even to employ a number of different spectrometers, each with different bandwidths, simultaneously. The second advantage follows directly from their digital nature. Once the signal is digitized, it is only mathematics. Tests on astronomical sources have shown that the noise follows a $1/\sqrt{Bt}$ behavior for integration times >100 hours; in these aspects, analogue spectrometers are more limited.

A serious drawback of digital auto and cross correlation spectrometers had been limited bandwidths. Previously 50–100 MHz had been the maximum possible bandwidth. This was determined by the requirement to meet Nyquist sampling rate, so that the analogue-to-digital (A/D) converters, samplers, shift registers and multipliers would have to run at a rate equal to twice the bandwidth. The speed of the electronic circuits was limited. However, advances in digital technology in recent years have allowed the construction of autocorrelation spectrometers with several 1000 channels covering bandwidths of several GHz.

Another improvement is the use of *recycling* auto and cross correlators. These spectrometers have the property that the product of bandwidth, B, times the number of channels, N, is a constant. Basically, this type of system functions have the digital part running at a high clock rate, while the data are sampled at a much slower rate. Then after the sample reaches the Nth shift register it is reinserted into the first register and another set of delays are correlated with the current sample. This leads to a higher number of channels and thus higher resolution. Such a system has the advantage of high-frequency resolution, but is limited in bandwidth. This has the greatest advantage for longer wavelength observations. Both of these developments have tended to make the use of digital spectrometers more widespread. This trend is likely to continue.

Autocorrelation systems are used in single telescopes, and make use of the symmetric nature of the autocorrelation function ACF. Thus, the number of delays gives the number of spectral channels. For cross-correlation, the current and delayed samples refer to different inputs. Cross-correlation systems are used in interferometers.

symmetric about zero time delay, but can be expressed in terms of amplitude and phase at each frequency, where both the phase and intensity of the line signal are unknown. Thus, for interferometry the zero delay of the ACF is placed in channel $N/2$ and is in general asymmetric. The number of delays, N, allows the determination of $N/2$ spectral intensities, and $N/2$ phases. The cross-correlation hardware can employ either an XF or a FX correlator. The FX correlator has the advantage that the time delay is just a phase shift, so can be introduced more simply.

4.6.1 Acousto-Optical Spectrometers

Since the discovery of molecular line radiation in the mm wavelength range there has been a need for spectrometers with bandwidths of several GHz. At 100 GHz, a velocity range of 300 km s^{-1} corresponds to 100 MHz, while the narrowest line widths observed correspond to 30 kHz. Autocorrelation spectrometers can reach such large bandwidths only if complex methods are used. The AOS makes use of the diffraction of light by ultrasonic waves: these cause periodic density variations in the medium through which it passes. These density variations in turn cause variations in the bulk constants ε and n of the medium, so that a plane electromagnetic wave passing through this medium will be affected. The most advanced AOS's have been designed and built at the First Physical Institute at Cologne University.

5 Filled Aperture Antennas

The material presented in the following emphasizes descriptive antenna parameters. These allow a fairly accurate but rather simple description of antenna properties that are needed for an accurate interpretation of astronomical measurements. For more detail, see [2]. An important concept is *reciprocity*, in which the properties of antennas are the same, irrespective whether these are used for receiving or transmitting. Reciprocity is limited under some (very special) technological applications that are not encountered in astronomy. In all of the following, only the far radiation field is considered.

5.1 Angular Resolution

From diffraction theory [23], the angular resolution of a reflector of diameter D at a wavelength λ is

$$\boxed{\theta = \frac{\lambda}{D}} \,. \tag{52}$$

This simple result gives a value for uniform illumination. This is the *best* that can be obtained. The next sections give some details of the characteristics of single dish antennas. The most important are calibration and efficiencies.

5.2 The Power Pattern $P(\vartheta, \varphi)$

Often, the *normalized power pattern*, not the power pattern is measured:

$$\boxed{P_{\mathrm{n}}(\vartheta, \varphi) = \frac{1}{P_{\max}} P(\vartheta, \varphi)} \, . \tag{53}$$

The reciprocity theorem provides a method to measure this quantity. The radiation source can be replaced by a small diameter radio source. The flux densities of such sources are determined by measurements using horn antennas at centimeter and millimeter wavelengths. At short wavelengths, one uses planets, or moons of planets, whose surface temperatures are determined from infrared data.

If the power pattern is measured using artificial transmitters, care should be taken that the distance from a large antenna A (diameter D) to a small antenna B (transmitter) is so large that B produces plane waves across the aperture D of antenna A, i.e., is in the far radiation field of A. This is the *Rayleigh* distance; it requires that the curvature of a wavefront emitted by B is much less than a wavelength across the geometric dimensions of A. This curvature must be $k \ll 2D^2/\lambda$, for an antenna of diameter D and a wavelength λ.

Consider the power pattern of the antenna used as a transmitter. If the spectral power density, P_ν in [W Hz^{-1}] is fed into a lossless isotropic antenna, this would transmit P power units per solid angle Ω per Hertz. Then the total radiated power at frequency ν is $4\pi\, P_\nu$. In a realistic, but still lossless antenna, a power $P(\vartheta, \varphi)$ per unit solid angle is radiated in the direction (ϑ, φ). If we define the directive gain $G(\vartheta, \varphi)$ as

$$P(\vartheta, \varphi) = G(\vartheta, \varphi)\, P$$

or

$$\boxed{G(\vartheta, \varphi) = \frac{4\pi P(\vartheta, \varphi)}{\iint P(\vartheta, \varphi)\, \mathrm{d}\Omega}} \, . \tag{54}$$

Thus the gain or directivity is also a normalized power pattern similar to Eq. (53), but with the difference that the normalizing factor is $\int P(\vartheta, \varphi)\, \mathrm{d}\Omega / 4\pi$. This is the gain relative to a lossless isotropic source. Since such an isotropic source cannot be realized in practice, a measurable quantity is the gain relative to some standard antenna such as a half-wave dipole whose directivity is known from theory.

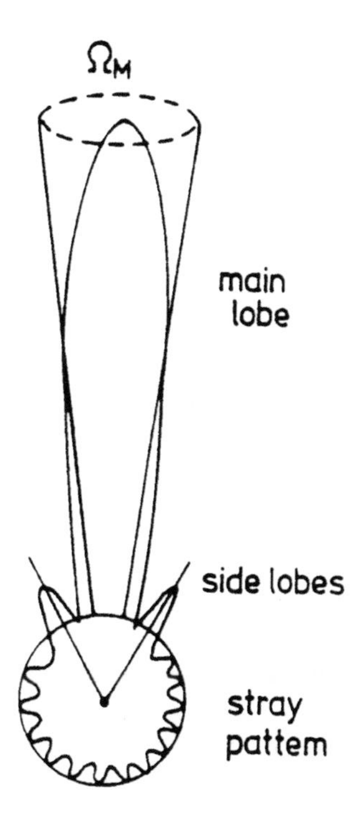

Fig. 5 A polar power pattern showing the main beam, and near and far side lobes. The weaker far side lobes have been combined to form the stray pattern

5.3 *The Main Beam Solid Angle*

The *beam solid angle* Ω_A of an antenna is given by

$$\boxed{\Omega_A = \iint\limits_{4\pi} P_n(\vartheta, \varphi)\, d\Omega = \int\limits_0^{2\pi}\int\limits_0^{\pi} P_n(\vartheta, \varphi) \sin\vartheta\, d\vartheta\, d\varphi}\,, \tag{55}$$

this is measured in steradians (sr). The integration is extended over the full sphere 4π, such that Ω_A is the solid angle of an ideal antenna having $P_n = 1$ for all of Ω_A and $P_n = 0$ everywhere else. Such an antenna does not exist; for most antennas the (normalized) power pattern has considerably larger values for a certain range of both ϑ and φ than for the remainder of the sphere. This range is called the main beam or main lobe of the antenna; the remainder are the side lobes or back lobes (Fig. 5). For actual situations, the properties are well defined up to the shortest operating wavelengths. At the shortest wavelength, there is still a main beam, but much of the power enters through side lobes. In addition, the main beam efficiency may vary significantly with elevation and weather has a large effect. Thus, the ability to accurately calibrate the radio telescope at sub-mm wavelengths is challenging.

In analogy to Eq. (55) we define the *main beam solid angle* Ω_{MB} by

$$\Omega_{\mathrm{MB}} = \iint\limits_{\substack{\text{main}\\ \text{lobe}}} P_{\mathrm{n}}(\vartheta, \varphi)\, \mathrm{d}\Omega \quad . \tag{56}$$

The quality of an antenna as a direction measuring device depends on how well the power pattern is concentrated in the main beam. If a large fraction of the received power comes from the side lobes it would be rather difficult to determine the location of the radiation source, the so-called "pointing".

It is appropriate to define a *main beam efficiency* or (in the slang of antenna specialists) *beam efficiency*, η_{B}, by

$$\eta_{\mathrm{B}} = \frac{\Omega_{\mathrm{MB}}}{\Omega_{\mathrm{A}}} \quad . \tag{57}$$

The main beam efficiency, η_{B}, is the fraction of the power concentrated in the main beam. The main beam efficiency can be modified (within certain limits) for parabolic antennas by the illumination of the main reflector. If the FWHP beamwidth is well defined, the location of an isolated source is determined to an accuracy given by the FWHP divided by the S/N ratio. Thus, it is possible to determine positions to small fractions of the FWHP beamwidth if noise is the only limit.

Substituting Eq. (55) into Eq. (54) it is easy to see that the maximum directive gain G_{max} or *directivity* $\mathcal{D}$ can be expressed as

$$\mathcal{D} = G_{\mathrm{max}} = \frac{4\pi}{\Omega_{\mathrm{A}}} \quad . \tag{58}$$

The angular extent of the main beam is usually described by the *half power beam width* (HPBW), which is the angle between points of the main beam where the normalized power pattern falls to $1/2$ of the maximum. For elliptically shaped main beams, values for widths in orthogonal directions are needed. The beamwidth is related to the geometric size of the antenna and the wavelength used; the exact beam size depends on details of illumination.

5.4 *Effective Area*

Let a plane wave with the power density $|\langle \boldsymbol{S} \rangle|$ be intercepted by an antenna. A certain amount of power is then extracted by the antenna from this wave; let this amount of power be P_{e}. We will then call the fraction

$$A_{\mathrm{e}} = P_{\mathrm{e}} \,/\, |\langle \boldsymbol{S} \rangle| \tag{59}$$

the *effective aperture* of the antenna. A_e has the dimension of m^2. Comparing this to the *geometric aperture* A_g we can define an aperture efficiency η_A by

$$\boxed{A_e = \eta_A A_g} . \tag{60}$$

Consider a receiving antenna with a normalized power pattern $P_n(\vartheta, \varphi)$ that is pointed at a brightness distribution $B_\nu(\vartheta, \varphi)$ in the sky. Then at the output terminals of the antenna, the total power per unit bandwidth, $\mathcal{P}_\nu$ is

$$\mathcal{P}_\nu = \tfrac{1}{2} A_e \iint B_\nu(\vartheta, \varphi) P_n(\vartheta, \varphi) \, d\Omega . \tag{61}$$

By definition, we are in the Rayleigh–Jeans limit, and can therefore exchange the brightness distribution by an equivalent distribution of brightness temperature. Using the Nyquist theorem (28) we can introduce an equivalent *antenna temperature* T_A by

$$\mathcal{P}_\nu = k T_A . \tag{62}$$

This definition of *antenna temperature* relates the output of the antenna to the power from a matched resistor. When these two power levels are equal, then the antenna temperature is given by the temperature of the resistor. Instead of the effective aperture A_e we can introduce the beam solid angle Ω_A. Then Eq. (61) becomes

$$\boxed{T_A(\vartheta_0, \varphi_0) = \frac{\int T_b(\vartheta, \varphi) P_n(\vartheta - \vartheta_0, \varphi - \varphi_0) \sin\vartheta \, d\vartheta \, d\varphi}{\int P_n(\vartheta, \varphi) \, d\Omega}} \tag{63}$$

which is the *convolution* of the brightness temperature with the beam pattern of the telescope. The brightness temperature $T_b(\vartheta, \varphi)$ corresponds to the thermodynamic temperature of the radiating material only for thermal radiation in the Rayleigh–Jeans limit from an optically thick source; in all other cases T_b is only a convenient quantity that in general depends on the frequency. It is important to note that from Eq. (63), the *measured* size of an extremely compact (i.e., "point") source is the beam size.

The quantity T_A in Eq. (63) was obtained for an antenna with no ohmic losses, and no absorption in the earth's atmosphere. In the mm/sub-mm range, the expression T_A in Eq. (63) is actually T_A', that is, a temperature corrected for atmospheric losses. We will use the term T_A' in discussions of mm/sub-mm calibration. Since T_A is the quantity measured while T_b is the one desired, Eq. (63) must be inverted. Equation (63) is an integral equation of the first kind, which in theory can be solved if the full range of $T_A(\vartheta, \varphi)$ and $P_n(\vartheta, \varphi)$ are known. In practice this inversion is possible only approximately. Usually both $T_A(\vartheta, \varphi)$ and $P_n(\vartheta, \varphi)$ are known only for a limited range of ϑ and φ values, and the measured data are not free of errors. Therefore, usually only an approximate deconvolution is performed. A special case is one for which the source distribution $T_b(\vartheta, \varphi)$ has a small extent compared to the telescope beam. Given a finite signal-to-noise ratio, the best estimate for the upper limit to the actual FWHP source size is one-half of the FWHP of the telescope beam. This point cannot be emphasized too much: we *cannot* assign an arbitrarily small size to a source. The best is one-half of the antenna FWHP!

5.5 Antenna Feed Horns Used Today

Feed horns are needed to guide the power from the reflector (in free space conditions) into the receiver (in a waveguide); details are contained in [14, 27]. The electric and magnetic field strengths at the open face of a wave guide will vary across the aperture. The power pattern of this radiation depends both on the dimension of the wave guide in units of the wavelength, λ, and on the mode of the wave. The greater the dimension of the wave guide in λ, the greater is the directivity of this power pattern. However, the larger the cross-section of a wave guide in terms of the wavelength, the more difficult it becomes to restrict the wave to a single mode. Thus wave guides of a given size can be used only for a limited frequency range. The aperture required for a selected directivity is then obtained by flaring the sides of a section of the wave guide so that the wave guide becomes a horn.

Great advances in the design of feeds have been made since 1960, and most parabolic dish antennas now use hybrid mode feeds (Fig. 6). Such "corrugated horns" are also referred to as *Scalar* or *Multi-Mode* feeds. Today such feed horns are used on all parabolic antennas. These provide much higher efficiencies than simple single mode horns and are well suited for polarization measurements.

Fig. 6 A corrugated waveguide hybrid mode feed for λ 2 mm on the Plateau de Bure interferometer. The one Euro coin in the upper left is shown to give a notion of scale. Photo courtesy of B. Lazareff (IRAM)

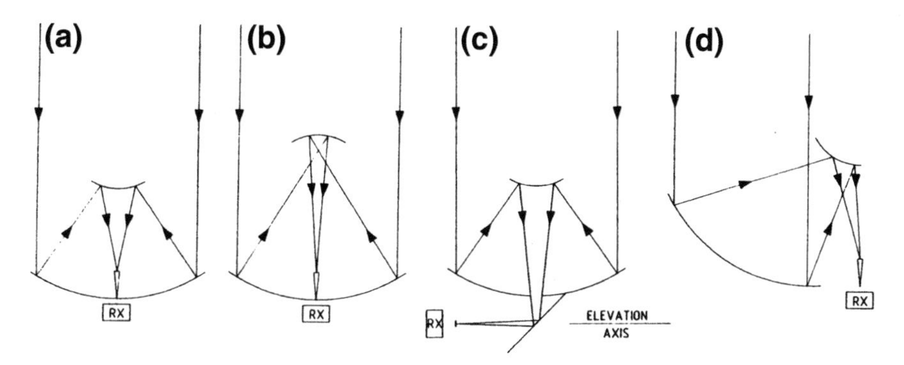

Fig. 7 The geometry of **a** Cassegrain, **b** Gregorian, **c** Nasmyth, and **d** offset Cassegrain systems. Figure taken from Rohlfs and Wilson (2004) [36]

5.6 Multiple Reflector Systems

If the size of a radio telescope is more than a few hundred wavelengths, designs similar to those of optical telescopes are preferred. For such telescopes Cassegrain, Gregorian and Nasmyth systems have been used (see Fig. 7). In a Cassegrain system, a convex hyperbolic reflector is introduced into the converging beam immediately in front of the prime focus. This reflector transfers the converging rays to a secondary focus which, in most practical systems is situated close to the apex of the main dish. A Gregorian system makes use of a concave reflector with an elliptical profile. This must be positioned behind the prime focus in the diverging beam. In the Nasmyth system this secondary focus is situated in the elevation axis of the telescope by introducing a flat mirror. The advantage of a Nasmyth system is that the receiver front ends remain horizontal while when the telescope is pointed toward different elevations. This is an advantage for receivers cooled with liquid helium, since these may become unstable when tipped. Cassegrain and Nasmyth foci are commonly used in the mm/sub-mm wavelength ranges.

In a secondary reflector system, feed illumination beyond the edge receives radiation from the sky, which has a temperature of only a few K. For low-noise systems, this results in only a small overall system noise temperature. This is significantly less than for prime focus systems. This is quantified in the so-called "G/T value", that is, the ratio of antenna gain of to system noise. Any telescope design must aim to minimize the excess noise at the receiver input while maximizing gain. For a specific antenna, this maximization involves the design of feeds and the choice of foci.

That the secondary reflector blocks the central parts in the main dish from reflecting the incoming radiation causes some interesting differences between the actual beam pattern from that of an unobstructed telescope. Modern designs seek to minimize blockage due to the support legs and sub-reflector.

Realistic filled aperture antennas ("single dishes") will have a beam pattern different from a uniformly illuminated unblocked aperture. First the illumination of the reflector will not be uniform but has a taper by 10 dB or more at the edge of the reflector. The side lobe level is strongly influenced by this taper: a larger taper lowers the side lobe level. Second, the secondary reflector must be supported by three or four support legs, which will produce aperture blocking and thus affect the shape of the beam pattern. In particular feed leg blockage will cause deviations from circular symmetry. For altitude-azimuth telescopes these side lobes will change position on the sky with hour angle. This may be a serious defect, since these effects will be significant for maps of low intensity regions if the main lobe is near an intense source. The side lobe response can also dependent on the polarization of the incoming radiation.

A disadvantage of on-axis systems, regardless of focus, is that they are often more susceptible to instrumental frequency baselines, so-called *baseline ripples* across the receiver band than primary focus systems. Part of this ripple is caused by multiple reflections of noise from source or receiver in the antenna structure. Ripples can arise in the receiver, but these can be removed or compensated rather easily. Telescope baseline ripples are more difficult to eliminate: it is known that large amounts of blockage and larger feed sizes lead to large baseline ripples. The effect is discussed in somewhat more detail in Sect. 6.4. The influence of baseline ripples on measurements can be reduced to a limited extent by appropriate observing procedures. A possible solution is the construction of off-axis systems such as the GBT.

5.7 Antenna Tolerance Theory

It is convenient to distinguish several different kinds of phase errors in the current distribution across the aperture of a two-dimensional antenna.

If the correlation distance d is of the same order of magnitude as the diameter of the reflector, part of the phase error can be treated as a systematic phase variation, either a linear error resulting only in a tilt of the main beam, or in a quadratic phase error which could be largely eliminated by refocussing. For $d \ll D$ the phase errors are almost independently distributed across the aperture, while for intermediate cases according to a good estimate for the expected value of the RMS phase error is given by:

$$\boxed{\bar{\delta^2} = \left(\frac{4\pi\varepsilon}{\lambda}\right)^2 \left[1 - \exp\left\{-\frac{\Delta^2}{d^2}\right\}\right]}, \tag{64}$$

where Δ is the distance between two points in the aperture to be compared, d is the correlation distance, and ε is the displacement of the reflector surface element in the direction of the wave. The gain of the system now depends both on $\bar{\delta^2}$ and on d. In addition, there is a complicated dependence both on the grading of the illumination and on the manner in which δ is distributed across the aperture. The Ruze theory can be expressed in the following terms: the gain of a reflector with surface phase errors

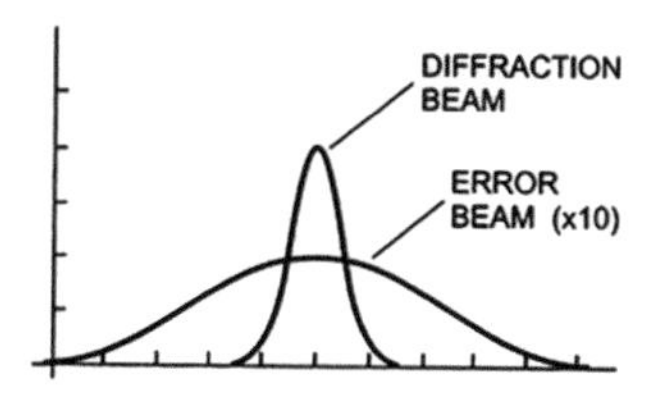

Fig. 8 A sketch showing the relative widths of the main beam and the error beam (shown ten times actual size). The width of the main beam is determined by diffraction for the entire antenna, while that for the error beam is determined by scale size of the surface irregularities. Figure taken from Wilson and Huettemeister (2005) [50]

can be approximated by the expression

$$G(u) = \eta\, e^{-\bar{\delta}^2} \left(\frac{\pi D}{\lambda}\right)^2 \Lambda_1^2\left(\frac{\pi D\, u}{\lambda}\right) + (1 - e^{-\bar{\delta}^2}) \left(\frac{2\pi d}{\lambda}\right)^2 \Lambda_1^2\left(\frac{2\pi d\, u}{\lambda}\right) , \qquad (65)$$

where

η is the aperture efficiency,
$u = \sin\vartheta$,
$\Lambda_1(u) = \frac{2}{u} J_1(u)$ is the Lambda function,
D the diameter of the reflector, and
d the correlation distance of the phase errors.

There are now two contributions to the beam shape of the system. The first is that of a circular aperture with a diameter D, but whose response is reduced due to the random phase error δ. The second term is the so-called *error beam* (see Fig. 8). This can be described as equal to the beam of a (circular) aperture with a diameter $2d$, its amplitude multiplied by

$$(1 - e^{-\bar{\delta}^2}) .$$

The error beam contribution therefore will decrease to zero as $\bar{\delta} \to 0$.

The gain of a filled aperture antenna with phase irregularities δ cannot increase indefinitely with increasing frequency but reaches a maximum at $\lambda_m = 4\pi\varepsilon$, and this gain is a factor of 2.7 below that of an error-free antenna of identical dimensions. Then, if the frequency can be determined at which the gain of a given antenna attains its maximum value, the RMS phase error and the surface irregularities ε can be measured electrically. Experience with many radio telescopes shows reasonably good agreement of such values for ε with direct measurements, giving empirical support for the Ruze tolerance theory.

6 Single Dish Observational Methods

6.1 The Earth's Atmosphere

For ground-based facilities, astronomical signals entering the receiver has been attenuated by the earth's atmosphere. In addition to attenuation, the receiver noise is increased by atmospheric emission, the signal is refracted and there are changes in the path length. Usually these effects change slowly with time, but there can also be rapid changes such as scintillation and anomalous refraction. Thus propagation properties must be taken into account, if the astronomical measurements are to be correctly interpreted. In the mm/sub-mm ranges, tropospheric effects are especially important. The various constituents of the atmosphere are absorbed by different amounts. Because the atmosphere can be considered to be in LTE, these constituents are also radio emitters.

The total amount of precipitable water (usually measured in mm) above an altitude h_0 is an integral along the line-of-sight. Frequently, the amount of H_2O is determined by measurements carried out at 225 GHz combined with models of the atmosphere. For mm/sub-mm sites, measurements of the 183 GHz spectral line of water vapor (see Fig. 21) can be used to estimate the total amount of H_2O in the atmosphere. For sea level sites, the 22.235 GHz line of water vapor is used for this purpose. The scale height $H_{H_2O} \approx 2$ km, is considerably less than $H_{air} \approx 8$ km of dry air. For this reason, sites for sub-millimeter radio telescopes are usually mountain sites with elevations above $\approx$3000 m.

The variation of the intensity of an extraterrestrial radio source due to propagation effects in the atmosphere is given by the standard relation for radiative transfer through a uniform medium (from Eq. (25))

$$\boxed{T_b(s) = T_b(0)\, e^{-\tau_\nu(s)} + T\,(1 - e^{-\tau_\nu(s)})}\,. \tag{66}$$

Here s is the (geometric) path length along the line-of-sight with $s = 0$ at the upper edge of the atmosphere and $s = s_0$ at the antenna. Both the (volume) absorption coefficient κ and the gas temperature T will vary with s, introducing the mass absorption coefficient k_ν by

$$\kappa_\nu = k_\nu \cdot \varrho\,, \tag{67}$$

where ϱ is the gas density; this variation of κ can be related to ϱ as long as the gas mixture remains constant along the line-of-sight. This is a simplified relation. For more realistic calculations, one must use a multi-layer model. Because the variation of ϱ with s is so much larger that that of $T(s)$, a useful approximation can be obtained by introducing an effective temperature for the atmosphere.

Refraction effects in the atmosphere depend on the real part of the (complex) index of refraction. Except for the anomalous dispersion near water vapor lines and oxygen lines, it is essentially independent of frequency. The average effect can be calculated; fine corrections are determined from pointing corrections.

A rapidly time variable effect is *anomalous refraction* (see [36]). If anomalous refraction is important, the apparent positions of radio sources appear to move away from their actual positions by up to $40''$ for time periods of 30 s. This effect occurs more frequently in summer afternoons than during the night. Anomalous refraction is caused by small variations in the H_2O content, perhaps by single cells of moist air. In the mm and sub-mm range, there are measurements of rapidly time variable noise contributions, the so-called *sky noise*. This is produced by variations in the water vapor content in the telescope beam. It does not depend in an obvious way on the transmission of the atmosphere. This behavior is expected if the effects arise within a few km above the telescope and the cells have limited sizes.

6.2 Millimeter and sub-mm Calibration Procedures

6.2.1 General

In radio astronomy, one usually follows a three step practical procedure: (1) the measurements must be corrected for atmospheric effects, (2) relative calibrations are made using secondary standards, and (3) if needed, gain versus elevation curves for the antenna must be established. In the mm/sub-mm ranges, primary calibrators are, in many cases, planets or moons of planets; more common secondary calibrators are non-time-variable compact sources.

6.2.2 Calibration of mm and sub-mm Wavelength Heterodyne Systems

In the mm/sub-mm wavelength range, the atmosphere has a larger influence and can change rapidly, so one must make accurate corrections to obtain well calibrated data. In the mm range, most large telescopes are close to the limits caused by their surface accuracy, so that the power received in the error beam may be comparable to that received in the main beam. Thus, one must use a relevant value of beam efficiency. We give an analysis of the calibration procedure which is standard in spectral line mm astronomy following the presentations in [10]. This calibration reference is referred to as the *chopper wheel* method. The procedure consists of: (1) the measurement of the receiver output when an ambient (room temperature) load is placed before the feed horn, and (2) the measurement of the receiver output, when the feed horn is directed toward cold sky at a certain elevation. For (1), the output of the receiver while measuring an ambient load, $T_{\rm amb}$, is $V_{\rm amb}$:

$$V_{\rm amb} = G\,(T_{\rm amb} + T_{\rm rx})\;. \tag{68}$$

For step (2), the load is removed; then the response to empty sky noise, $T_{\rm sky}$, and receiver cabin (or ground), $T_{\rm gr}$, is

$$V_{\rm sky} = G\,[F_{\rm eff}\,T_{\rm sky} + (1 - F_{\rm eff})\,T_{\rm gr} + T_{\rm rx}]\,. \tag{69}$$

$F_{\rm eff}$ is referred to as the *forward efficiency*. This is basically the fraction of power in the forward beam of the feed. If we take the difference of $V_{\rm amb}$ and $V_{\rm sky}$, we have

$$V_{\rm cal} = V_{\rm amb} - V_{\rm sky} = G\,F_{\rm eff}\,T_{\rm amb}\,{\rm e}^{-\tau_\nu}\,, \tag{70}$$

where τ_ν is the atmospheric absorption at the frequency of interest. The response to the signal received from the radio source, through the earth's atmosphere, is

$$\Delta V_{\rm sig} = G\,T'_{\rm A}\,{\rm e}^{-\tau_\nu}$$

or

$$T'_{\rm A} = \frac{\Delta V_{\rm sig}}{\Delta V_{\rm cal}}\,F_{\rm eff}\,T_{\rm amb}\,,$$

where $T'_{\rm A}$ is the antenna temperature of the source outside the earth's atmosphere. We define

$$T^*_{\rm A} = \frac{T'_{\rm A}}{F_{\rm eff}} = \frac{\Delta V_{\rm sig}}{\Delta V_{\rm cal}}\,T_{\rm amb}\,. \tag{71}$$

The quantity $T^*_{\rm A}$ is commonly referred to as the *corrected antenna temperature*, but it is really a *forward beam brightness temperature*. This is the $T_{\rm MB}$ of a source filling a large part of the sky, certainly more than 30′.

For sources (small compared to 30′), one must still correct for the telescope beam efficiency, which is commonly referred to as $B_{\rm eff}$. Then

$$T_{\rm MB} = \frac{F_{\rm eff}}{B_{\rm eff}}\,T^*_{\rm A}$$

for the IRAM 30 m telescope, $F_{\rm eff} \cong 0.9$ down to 1 mm wavelength, but $B_{\rm eff}$ varies with the wavelength. So at $\lambda = 3$ mm, $B_{\rm eff} = 0.65$, at 2 mm $B_{\rm eff} = 0.6$ and at 1.3 mm $B_{\rm eff} = 0.45$, for sources of diameter $< 2'$. For an object of size 30′, $B_{\rm eff}$ at all these wavelengths is 0.65. As usual $T_{\rm MB}$ can be considered a black body with the temperature $T_{\rm MB}$, which just fills the beam. This analysis is the one used at IRAM.

In terms of our notation

$$\eta_{\rm MB} = \frac{\Omega_{\rm MB}}{\Omega_{\rm F}} = \frac{B_{\rm eff}}{F_{\rm eff}}\,.$$

An antenna pointing at an elevation z to a position of empty sky will deliver an antenna temperature

$$T_{\rm A}(z) = T_{\rm rx} + T_{\rm atm}\,\eta_{\rm l}\,(1 - {\rm e}^{-\tau_0 X(z)}) + T_{\rm amb}(1 - \eta_{\rm l})\;, \tag{72}$$

where

$T_{\rm rx}$: system noise temperature,
$T_{\rm atm}$: effective temperature of the atmosphere,
$T_{\rm amb}$: ambient temperature,
$\eta_{\rm l}$: feed efficiency (typically $\eta_{\rm l} = 0.9$),
τ_0: zenith optical depth, and
$X(z)$: air mass at zenith distance z.

These parameters can be determined by a series of calibration measurements. The efficiency $\eta_{\rm l}$ and the other parameters can be determined by a least squares fit of Eq. (72), that is a *skydip* giving $T_{\rm A}$ as a function of $X(z)$. Depending on the weather conditions these measurements have to be repeated at time intervals from 15 min to hours or so, to be able to detect variations in the atmospheric conditions. At some observatories a small separate instrument, a *taumeter* (a sky horn that measures the sky temperature at elevations 90°, 60°, 30° and 20°) is available to determine the opacity τ at 10 min intervals.

For larger mm wavelength telescopes one cannot perform tipping measurement often. If a taumeter is not available one must use a more elaborate procedure. By measuring the response to a cold load, one can determine the receiver noise, and can obtain a good estimate of the noise from the atmosphere. Then, assuming a value of $T_{\rm atm}$ and $\eta_{\rm l} = F_{\rm eff}$, one can then determine $\tau = \tau_0\,X(z)$, and can use this to correct for atmospheric absorption.

To calibrate spectral lines, one frequently measures sources for which one has single sideband spectra. Finally observations often have to be corrected for yet another effect: the telescope efficiency usually depends on elevation. Usually the telescope surface is set optimally for some intermediate zenith distance $z \approx 40°$. Both for $z \approx 0°$ and 70° the efficiency usually decreases somewhat.

6.2.3 Bolometer Calibrations

Since most bolometers are A. C. coupled (i.e., responds to differences), the D. C. response (i.e., responds to total power) to "hot-cold" or "chopper wheel" calibration methods are not used. Instead astronomical data are calibrated in two steps: (1) measurements of atmospheric emission to determine the opacities at the azimuth of the target source, and (2) the measurement of the response of a nearby source with a known flux density; immediately after this, a measurement of the target source is carried out.

6.2.4 Compact Sources

Usually the beam of radio telescopes are well characterized by Gaussians, which have the great advantage that the convolution of two Gaussians is another Gaussian. For Gaussians, the relation between the observed source size, θ_o, the beam size θ_b, and actual source size, θ_s, is given by:

$$\theta_\mathrm{o}^2 = \theta_\mathrm{s}^2 + \theta_\mathrm{b}^2 \,. \tag{73}$$

This is a completely general relation, and is widely used to deconvolve source from beam sizes. Even when the source shapes are not well represented by Gaussians these are usually approximated by sums of Gaussians in order to have a convenient representation of the data. The accuracy of any determination of source size is limited by Eq. (73). A source whose size is less than 0.5 of the beam is so uncertain that one can only give as an upper limit of $0.5\,\theta_\mathrm{b}$.

If the (lossless) antenna (outside the earth's atmosphere) is pointed at a source of known flux density S_ν with an angular diameter that is small compared to the telescope beam, a power $W_\nu\,\mathrm{d}\nu$ at the receiver input terminals

$$W_\nu\,\mathrm{d}\nu = \tfrac{1}{2} A_\mathrm{e}\,S_\nu\,\mathrm{d}\nu = k\,T_\mathrm{A}'\,\mathrm{d}\nu$$

is available. Here T_A' is the antenna temperature corrected for effect of the earth's atmosphere. Thus

$$\boxed{T_\mathrm{A}' = \Gamma S_\nu} \,, \tag{74}$$

where Γ is the *sensitivity* of the telescope measured in K Jy^{-1}. Introducing the aperture efficiency η_A according to Eq. (60) we find

$$\boxed{\Gamma = \eta_\mathrm{A}\frac{\pi D^2}{8k}} \,. \tag{75}$$

Thus Γ or η_A can be measured with the help of a calibrating source provided that the diameter D and the noise power scale in the receiving system are known. In practical work the inverse of equation (74) is often used. Inserting numerical values we find

$$S_\nu = 3520\,\frac{T_\mathrm{A}'[\mathrm{K}]}{\eta_\mathrm{A}[D/\mathrm{m}]^2} \,. \tag{76}$$

The *brightness temperature* is defined as the Rayleigh–Jeans temperature of an equivalent black body which will give the same power per unit area per unit frequency interval per unit solid angle as the celestial source. Both T_A' and T_MB are defined in the classical limit, and *not* through the Planck relation. However the brightness temperature scale has been corrected for antenna efficiency. The conversion from source flux density to source brightness temperature for sources with sizes small compared to the telescope beam is given by Eq. (22). For sources small compared to the beam, the antenna and main beam brightness temperatures are related by the main beam efficiency η_B:

$$\eta_B = \frac{T'_A}{T_{MB}} . \tag{77}$$

The actual source brightness temperature, T_s, is related to the main beam brightness temperature by:

$$T_s = T_{MB} \frac{(\theta_s^2 + \theta_b^2)}{\theta_s^2} , \tag{78}$$

where we have made the assumption that source and beam are Gaussian shaped. The actual brightness temperature is a property of the source. To determine T_s, one must determine the actual source size (i.e., the "resolved" source), so that the source is extended compared to the beam (θ_s larger than θ_b). This is one of the major science drivers for high angular resolution (i.e., interferometry) measurements. Although the source may not be Gaussian shaped, one normally fits multiple Gaussians to obtain the effective source size.

6.2.5 Extended Sources

For sources extended with respect to the beam, the process is vastly more complex, because the antenna side lobes also receive power from the celestial source, and a simple relation using beam efficiency is not possible without detailed measurements of the antenna pattern. The error beam may be a very significant source of calibration errors, particularly if the measurements are carried out near the limit of telescope surface accuracy. In principle η_{MB} could be computed by numerical integration of $P_n(\vartheta, \varphi)$ (cf. Eqs. (55) and (56)), provided that $P_n(\vartheta, \varphi)$ could be measured for large range of ϑ and φ. Unfortunately this is not possible since nearly all astronomical sources are too weak; measurements of bright astronomical objects with known diameters can be useful.

If we assume a source has a uniform brightness temperature over a certain solid angle Ω_s, then the telescope measures an antenna temperature given by Eq. (63) which, for a constant brightness temperature across the source, simplifies to

$$T'_A = \frac{\displaystyle\int_{\text{source}} P_n(\vartheta, \varphi)\, d\Omega}{\displaystyle\int_{4\pi} P_n(\vartheta, \varphi)\, d\Omega} T_b$$

or, introducing Eqs. (55)–(57),

$$T'_A = \eta_B \frac{\int_{\text{source}} P_n(\vartheta, \varphi)\, d\Omega}{\int_{\text{main lobe}} P_n(\vartheta, \varphi)\, d\Omega} T_b = \eta_B f_{\text{BEAM}} T_b\,, \tag{79}$$

where f_{BEAM} is the beam filling factor. For Gaussians

$$f_{\text{BEAM}} = \frac{\theta_s^2}{(\theta_s^2 + \theta_b^2)}$$

if the source diameter is of the same order of magnitude as the main beam the correction factor in Eq. (79) can be determined with high precision from measurements of the normalized power pattern and thus Eq. (79) gives a direct determination of η_B, the beam efficiency. A convenient source with constant surface brightness in the long cm wavelength range is the moon whose diameter of $\cong 30'$ is of the same order of magnitude as the beams of most large radio telescopes and whose brightness temperature

$$T_{\text{b moon}} \cong 225\ \text{K} \tag{80}$$

is of convenient magnitude. In the mm and sub-millimeter range the observed Moon temperature changes with Moon phase. The planets form convenient thermal sources with known diameters that can be used for calibration purposes [1, 39].

6.3 Continuum Observing Strategies

6.3.1 Point Sources

In the sub-mm range the earth's atmosphere is a large source of radiation. Compensation of transmission variations in the atmosphere is possible if double beam systems can be used. At higher frequencies, in the mm/sub-mm range, the rapid movement of the telescope beam (by small movements of the sub-reflector or a mirror in the path from receiver to antenna) over small angles, so-called "wobbling" is used to produce two beams on the sky from a single pixel. This is used at all large millimeter facilities. In the simplest system the individual telescope beams should be spaced by a distance of at least 3 FWHP beam widths, and the receiver should be switched between them. The separate beams can be implemented in different ways depending on the frequency and the technical facilities at the telescope. Observing procedures for a double beam system are usually as follows: the source is first centred on beam one, and the difference of the two beams is measured, optimally by wobbling the sub-reflector. Then the source is centred on beam two, and again the difference is measured. This on–off method (better called on–on, because the source is always in one of the beams) is often arranged in a time symmetric fashion so that time variations of the sky noise and other instrumental effects can be eliminated.

Multi-beam bolometer systems are now the rule. With these, one can measure a fairly large region simultaneously. This allows a higher mapping speed, and also provides a method to better cancel sky noise due to weather. Such weather effects are referred to as "coherent noise". Some details of more recent data methods are given in e.g. [33]. Usually, a wobbler system is needed for such arrays, since the bolometer outputs are usually A. C.–coupled.

6.3.2 Imaging of Extended Continuum Sources

If extended areas are to be mapped, some kind of raster scan is employed: there must be reference positions at the beginning and the end of the scan. Usually the area is measured at least twice in orthogonal directions. After gridding, the differences of the images are least squares minimized to produce the best result. This procedure is called "basket weaving".

Extended emission regions can also be mapped using a double beam system, with the receiver input periodically switched between the first and second beam. In this procedure, there is some suppression of very extended emission. A simple summation along the scan direction has been used to reconstruct infrared images. More sophisticated schemes can recover most, but not all of the information [33]. Most telescopes employ wobbler switching in azimuth to cancel ground radiation. By measuring a source using scans in azimuth at different hour angles, and then combining the maps one can recover more information and thus a more complete determination of the source structure [24].

6.4 Additional Requirements for Spectral Line Observations

In addition to the requirements placed on continuum receivers, there are three requirements specific to spectral line receiver systems.

6.4.1 Radial Velocity Settings

If the observed frequency of a line is compared to the known rest frequency, the relative radial velocity of the line emitting (or absorbing) source and the receiving system can be determined. But this velocity contains the motion of the source as well as that of the receiving system. Both are measured relative to some standard of rest. However, usually only the motion of the source is of interest. Thus the velocity of the receiving system must be determined. This velocity can be separated into several independent components: **(1) Earth Rotation** with a maximum velocity $v = 0.46$ km s^{-1} and **(2) The Motion of the Center of the Earth** relative to the barycenter of the Solar System is said to be reduced to the *heliocentric* system. Correction algorithms are available for observations of the earth relative to center of mass of the solar system. The resulting radial velocities are then as close to an inertial system. Results obtained by many independent investigations show that the solar system moves with a velocity given by the *standard solar motion*. This is the solar motion relative to the mode of the velocity of the stars in the solar neighborhood. Data from which the standard solar motion has been eliminated are said to refer to the *local standard of rest* (LSR).

6.4.2 Stability of the Frequency Bandpass

In addition to the stability of the total power of the receiver, one must also have a stable shape of the receiver bandpass. At millimeter and sub-mm wavelengths, it is possible that changes in the weather conditions between on-source and reference measurements may lead to serious baseline instabilities. If so, the time between on-source and reference measurements must be shortened until stable conditions are reached. Such stability is easier to obtain if the bandwidth of the spectrometer is narrow compared to the bandwidth of those parts of the receiver in front of the spectrometer.

6.4.3 Instrumental Frequency Effects

The result of any observing procedure should result in a spectrum in which $T_{\mathrm{A}}(\nu) \rightarrow 0$ for ν outside the frequency range of the line. However, quite often this is not so because the signal response was not completely compensated for by a reference measurement, even if receiver stability is ideal. For larger bandwidths, there is an instrumental spectrum and a "baseline" must be subtracted from the difference spectrum. Often a linear function of frequency is sufficient, but sometimes some curvature is found, so that polynomials of second or higher order must be subtracted.

Often a sinusoidal or quasi-periodic baseline ripple is present because a small fraction of the signal is reflected off obstructions in the antenna. This reflected signal can form a standing wave pattern. A phase change of 2π radians will occur if either the distance, d, over which the signals are interfering is changed by $\lambda/2$ (where λ is the wavelength) or if the frequency is changed by

$$\Delta\nu = \frac{c}{2d} \,. \tag{81}$$

There are several possible sources of reflected radiation: (1) the receiver that injects some noise power into the antenna, part of which is then reflected back; or (2) strong continuum radiation from sources or the atmosphere. In both cases the partial reflection of the radiation in the horn aperture is the main cause of the instrumental "baseline ripples". Both changes in the position of the telescope and small changes in the receiving equipment can cause large changes in the amplitude of the observed ripple. Sometimes the amplitude of baseline ripple can be reduced considerably by installing a cone at the apex of the telescope that scatters the radiation forming the standing wave pattern.

6.4.4 Spectral Line Observing Strategies

In radio astronomy spectral line radiation is almost always only a small fraction of the total power received; the signal sits on a large pedestal of wide band noise signals contributed by different sources: the system noise, spillover from the antenna and in some cases, a true background noise. To avoid the stability problems encountered in total power systems the signal of interest must be compared with another signal that contains the same total power and differs from the first only in that it contains no line radiation. To achieve this aim, mm/sub-mm spectral line observers usually make use of three observing modes that differ only in the way the comparison signal is produced.

(1) Switching Against an Absorber. Today this method is used only in exceptional circumstances such as for some studies of the 2.7 K cosmic microwave background.

(2) Frequency Switching. For many sources, spectral line radiation is a narrow-band feature, that is, the emission is centred at ν_0, present over a small frequency interval, $\Delta\nu$, with $\nu_0/\Delta\nu \approx 10^{-6}$. If all other effects vary very little over $\Delta\nu$, changing the frequency of a receiver by perhaps 10 $\Delta\nu$ produces a comparison signal with the line shifted. If other contributions hardly differ, the final spectrum is proportional to the difference of these two measurements. Such "frequency switched" measurements can be done with almost any rate. These produce a particularly good compensation for wide-band atmospheric instabilities. Such observations can be made for mm wave radiation even in poor weather conditions but functions best for lines having widths of less than a few MHz. If the spectral line is included in the analysing band in both the signal and the reference phases, the effective integration time is doubled.

(3) Position Switching and Wobbler Switching. The received signal "on source" is compared with another signal obtained at a nearby position in the sky. If the emission is rather extended and the atmospheric effects are large (for example in the case of galactic Carbon Monoxide emission), one may need to use two reference measurements, one at a higher, and the other at a lower elevation. A number of conditions must be fulfilled: (1) the receiver is stable so that any gain and bandpass changes occur only over time scales which are long compared to the time needed for position change, and (2) there is little line radiation at the comparison region. If so, this method is efficient and produces excellent line profiles. A variant of this method is wobbler switching. This is very useful for compact sources, especially in the mm and sub-mm range.

(4) On the Fly Mapping. This very important observing method is an extension of method (3). In this procedure, one takes spectral line data at a rate of perhaps one spectrum or more per second. As with total power observing, usually one first takes a reference spectrum, and then takes data along a given direction. Then one changes the position of the telescope in the perpendicular direction, and repeats the procedure until the entire region is sampled. Because of the short integration times an entire image of perhaps $15' \times 15'$ taken with a $30''$ beam could be finished in roughly 20 min. At each position, the S/N ratio may be low, but the procedure can be repeated. With each data transfer, the telescope position is read out. Even if there are absolute pointing errors, over this short time and small angle the relative positions where spectra were taken are accurate. The resulting accuracy is improved because the spectra are oversampled and weather conditions are more uniform over the region mapped. To produce the final image the individual spectra are placed on a grid and then averaged.

7 Interferometers and Aperture Synthesis

From diffraction theory Eq. (52), the angular resolution of a radio telescope is $\theta = k\,\lambda/D$, where θ is the angular resolution, λ is the wavelength of the radiation received, D is the diameter of the instrument and k is a factor of order unity. For a given wavelength, the diameter D must be increased to improve this angular resolution. The largest mm/sub-mm antenna is the Large Millimeter Telescope (LMT), with a 50 meter diameter, and a short wavelength limit of 0.8 mm, so $\theta \cong 3''$. An alternative method was proposed and put into practice by Michelson [23]. In this, a resolving power $\theta \approx \lambda/D$ can be obtained by coherently combining the output of two reflectors of diameter $d \ll D$ separated by a distance D.

An important extension of interferometry is aperture synthesis, that is, the production of high quality images of a source by combining a number of independent measurements in which the antenna spacings cover an equivalent aperture. In this chapter, we give an introduction to the principles of aperture synthesis. Vastly more detailed accounts with lots and lots of math are to be found in [46] or [11].

7.1 Two Element Interferometers

The basic principles can be understood from a consideration of Fig. 9. In panel (a) is the response of a single uniformly illuminated aperture of diameter D. In panels (b) and (c) we show the response of a two element interferometer consisting of two small antennas (diameter d) separated by a distance D and $2D$, where $d \ll D$. The interferometer response is obtained from the multiplication of the outputs of the two antennas. The uniformly illuminated aperture has a dominant main beam of width $\theta = k\,\lambda/D$, accompanied by smaller secondary maxima, or side lobes. There are two differences between the case of a single dish response compared to the case of an interferometer. First, for an interferometer, astronomers use a different nomenclature.

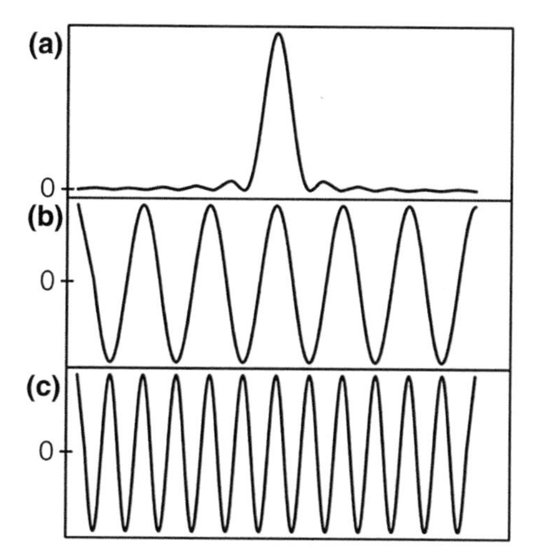

Fig. 9 Power patterns for different antenna configurations. The horizontal axis in this figure is angle. Panel **a** shows that of a uniformly illuminated full aperture with a full width to half power (FWHP) of $k\,\lambda/D$, with $k \approx 1$. In panel **b** we show the power pattern of a two element multiplying interferometer consisting of two antennas of diameter d spaced by a distance D where $d \ll D$. In panel **c** we show the power pattern of the interferometer system described in **b** but now with a spacing $2D$

Instead of "main beam and side lobes" one speaks of "fringes". There is a central fringe (or "white light" fringe in the analogy with Young's Two Slit experiment) and adjacent fringes. Second, we will show that with the multiplication of the outputs of two antennas, the fringes are centred about zero so that the total power output of each antenna is suppressed. Since some of the information (i.e., total power) is not used, for a given spacing only source structure comparable to (or smaller than) a fringe is recorded fully. For the case of an interferometer composed of two small dishes (with dish diameter $d \ll D$) there is no prominent main beam and the side lobe level does not markedly decrease with increasing angular offset from the axes of the antennas. Comparing the width of the fringes in panels (b) and (c) one finds that by doubling the separation D of the antennas, the width of the fringe width is halved. For the interferometer spacing (usually referred to as *the baseline*) D, in panel (b) the resolving power of the filled aperture is not greatly different from the single dish in panel (a), but the collecting area of this two element interferometer is smaller. For larger spacings, the interferometer angular resolution is greater.

If a uniform source is extended in angle by a positive and a negative fringe in Fig. 9 the response of the multiplied output is zero. For source structure smaller than a fringe, the response is not diminished. Thus by increasing D, finer and finer source structure is measured. Combining the outputs of independent data sets for spacings of D and $2D$ shows that these select different structural components of the source. Finer source structure can be recorded if in addition, $D\,n$ antenna spacings are measured. Such a series measurements can be made by increasing the separation of two antennas whose outputs are coherently combined.

A general procedure, *aperture synthesis*, is now the standard method to obtain high quality, high angular resolution images. The first practical demonstration of aperture synthesis in radio astronomy was made by Ryle and his associates. Aperture synthesis allows us to reproduce the imaging properties of a large aperture by sampling the radiation field at individual positions within the area contained within the aperture. In analogy with the approach used by Michelson in the optical wavelength range, the advance in radio astronomy was to measure the *mutual coherence function* and to show that these results were sufficient to produce images. Using this approach, a remarkable improvement of the radio astronomical imaging was possible.

Electromagnetic waves induce the voltage V_1 at the output of antenna A_1

$$V_1 \propto E\,\mathrm{e}^{\,\mathrm{i}\omega t}\,, \tag{82}$$

while at A_2 we obtain

$$V_2 \propto E\,\mathrm{e}^{\,\mathrm{i}\omega\,(t-\tau)}\,, \tag{83}$$

where E is the amplitude of the incoming plane wave, τ is the geometric delay caused by the relative orientation of the interferometer baseline $\boldsymbol{B}$ and the direction of the wave propagation. For simplicity, in Eqs. (82) and (83) we have neglected receiver noise and instrumental phase. The outputs will be correlated. Today, all interferometers use direct correlation, since the goal is to measure the correlation accurately. In a *correlation* the signals are input to a multiplying device followed by an integrator. The response of the correlator, R, is proportional to

$$R(\tau) \propto \frac{E^2}{T}\int_0^T \mathrm{e}^{\,\mathrm{i}\omega t}\,\mathrm{e}^{\,-\mathrm{i}\omega(t-\tau)}\,\mathrm{d}t\,.$$

If T is a time much longer than the time of a single full oscillation, i.e., $T \gg 2\pi/\omega$ then the average over time T will not differ much from the average over a single full period; that is

$$\begin{aligned} R(\tau) &\propto \frac{\omega}{2\pi}E^2 \int_0^{2\pi/\omega} \mathrm{e}^{\,\mathrm{i}\omega\tau}\,\mathrm{d}t \\ &\propto \frac{\omega}{2\pi}E^2\,\mathrm{e}^{\,\mathrm{i}\omega\tau}\int_0^{2\pi/\omega}\mathrm{d}t\,, \end{aligned}$$

resulting in

$$\boxed{R(\tau) \propto \tfrac{1}{2}E^2\,\mathrm{e}^{\,\mathrm{i}\omega\tau}}\,. \tag{84}$$

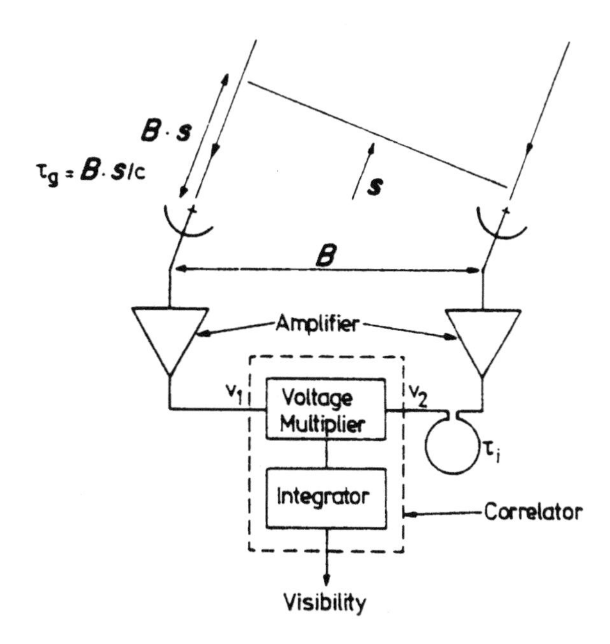

Fig. 10 A schematic diagram of a two-element correlation interferometer. The antenna output voltages are V_1 and V_2; the instrumental delay is $\tau_{\rm i}$ and the geometric delay is $\tau_{\rm g}$. Figure taken from Rohlfs and Wilson (2004) [36]

The output of the correlator + integrator thus varies periodically with τ, the delay time. If the orientation of interferometer baseline $\boldsymbol{B}$ and wave propagation direction $\boldsymbol{s}$ remain invariable, τ remains constant, so does $R(\tau)$. But since $\boldsymbol{s}$ is slowly changing due to the rotation of the earth, τ will vary, and we will measure *interference fringes* as a function of time.

In order to understand the response of interferometers in terms of measurable quantities, we consider a two-element system. The basic constituents are shown in Fig. 10. If the radio brightness distribution is given by $I_\nu(\boldsymbol{s})$, the power received per bandwidth $\mathrm{d}\nu$ from the source element $\mathrm{d}\Omega$ is $A(\boldsymbol{s})I_\nu(\boldsymbol{s})\,\mathrm{d}\Omega\,\mathrm{d}\nu$, where $A(\boldsymbol{s})$ is the effective collecting area in the direction $\boldsymbol{s}$; we will assume the same $A(\boldsymbol{s})$ for each of the antennas. The amplifiers are assumed to have constant gain and phase factors (neglected for simplicity).

The output of the correlator for radiation from the direction $\boldsymbol{s}$ (Fig. 10) is

$$r_{12} = A(\boldsymbol{s})\,I_\nu(\boldsymbol{s})\;\mathrm{e}^{\,\mathrm{i}\omega\tau}\;\mathrm{d}\Omega\,\mathrm{d}\nu\,, \tag{85}$$

where τ is the difference between the geometrical and instrumental delays $\tau_{\rm g}$ and $\tau_{\rm i}$. If $\boldsymbol{B}$ is the baseline vector for the two antennas

$$\tau = \tau_{\rm g} - \tau_{\rm i} = \frac{1}{c}\boldsymbol{B}\cdot\boldsymbol{s} - \tau_{\rm i} \tag{86}$$

and the total response is obtained by integrating over the source S

$$R(\boldsymbol{B}) = \iint\limits_{\Omega} A(\boldsymbol{s}) I_\nu(\boldsymbol{s}) \exp\left[\mathrm{i}\, 2\pi\nu\left(\frac{1}{c}\boldsymbol{B}\cdot\boldsymbol{s} - \tau_\mathrm{i}\right)\right] \mathrm{d}\Omega\, \mathrm{d}\nu \quad . \tag{87}$$

This function $R(\boldsymbol{B})$, the *Visibility Function* is closely related to the mutual coherence function of the source, except for the power pattern $A(\boldsymbol{s})$ of the individual antennas. For parabolic antennas it is usually assumed that $A(\boldsymbol{s}) = 0$ outside the main beam area so that Eq. (87) is integrated only over this region. A one dimensional version of Eq. (87), with a baseline B, $\nu = \nu_0$ and $\tau_i = 0$, is

$$R(B) = \int A(\theta)\, I_\nu(\theta) \exp\left[\mathrm{i}\, 2\pi\nu_0\left(\frac{1}{c}B\cdot\theta\right)\right] \mathrm{d}\theta\ . \tag{88}$$

Such a one dimensional relation is a very useful guide to understand interferometer responses.

7.1.1 Calibration

Two quantities that must be calibrated for continuum measurements are amplitude and phase. In addition, for spectral line measurements the instrument passband must also calibrated.

The amplitude scale is calibrated using methods that are similar to those used for single dish measurements. This consists of using the response of each antenna to determine the system noise of the receiver being used. In the centimeter range, the atmosphere plays a small role while in the millimeter and sub-mm wavelength ranges, the atmospheric effects must be accounted for. For phase measurements, a suitable point-like source with an accurately known position is required to determine the instrumental phase $2\pi\nu\tau_i$ in Eq. (87). For interferometers, the calibration sources are usually unresolved or point-like sources. Most often these are extragalactic time variable sources. To calibrate the response in units of flux density or brightness temperatures, these measurements must be referenced to a thermal calibrator.

The calibration of the instrument passband is carried out by an integration of an intense source to determine the channel-to-channel gains and offsets. The amplitude, phase and passband calibrations are carried out before the source measurements. The passband calibration is usually carried out once per observing session. The amplitude and phase calibrations are made more often. Their frequency depends on the stability of the electronics and weather. At millimeter wavelengths, the calibrations are usually made every few minutes, but may have to be made more often in worse weather or at shorter wavelengths. If weather demands that frequent measurements of calibrators are required, this is referred to as *fast switching*.

7.2 *Responses of Interferometers*

7.2.1 Finite Bandwidth

Equation (87) can be used to estimate the effect of a finite bandwidth $\Delta\nu$. The geometric delay $\tau_g = \frac{1}{c}\boldsymbol{B}\cdot\boldsymbol{s}$ is by definition independent of frequency, but the instrumental delay τ_i may not be. Adjusting τ_i the sum $\tau = \tau_g - \tau_i$ can be made equal to zero for the center of the band. Introducing the relative phase of a wave by

$$\varphi = \left[\frac{c\tau}{\lambda}\right]_{\text{fractional part}} ,$$

we obtain

$$\varphi = \frac{1}{\lambda}\boldsymbol{B}\cdot\boldsymbol{s} + \varphi_i , \quad (89)$$

where φ_i is the instrumental phase corresponding to the instrumental delay. This phase difference varies across the band of the interferometer $\Delta\nu$ by

$$\Delta\varphi = \frac{1}{\lambda}\boldsymbol{B}\cdot\boldsymbol{s}\,\frac{\Delta\nu}{\nu} . \quad (90)$$

The fringes will disappear when $\Delta\varphi \simeq 1$ radian. As can be seen the response is reduced if the frequency range, that is, the bandwidth, is large compared to the delay caused by the separation of the antennas. For large bandwidths, the loss of visibility can be minimized by adjusting the phase delay, the time difference is negligible (see Fig. 10). In effect, this is only possible if the exponential term in Eq. (87) is kept small. In practice, this is done by inserting a delay between the antennas so that $\frac{1}{c}\boldsymbol{B}\cdot\boldsymbol{s}$ equals τ_i. In the first interferometric systems this was done by switching lengths of cable into the system; currently this is accomplished by first digitizing the signal after conversion to an intermediate frequency, and then using digital shift registers. In analogy with the optical wavelength range, this adjustment of cable length is equivalent to centring the response on the central, or *white light fringe* in Young's two-slit experiment.

The reduction of the response caused by finite bandwidth can be estimated by an integration of Eq. (87) over frequency. Taking $A(\boldsymbol{s})$ and $I_\nu(\boldsymbol{s})$ to be constants, and integrating over a range of frequencies $\Delta\nu = \nu_1 - \nu_2$. Then the result is an additional factor, $\sin(\Delta\nu\tau)/\Delta\nu\tau$ in Eq. (87). This will reduce the interferometer response if $\Delta\varphi \sim 1$. For typical bandwidths of 100 MHz, the offset from the zero delay must be $\ll 10^{-8}$ s. This adjustment of delays is referred to as *fringe stopping*. This causes the response of Eq. (87) to lose a component. To recover this input, an extra delay of a quarter wavelength relative to the input of the correlator is inserted, so that the sine and cosine response in Eq. (87) can be measured. In digital cross-correlators, (see Sect. 4.6), the sine and cosine components are obtained from the positive and negative delays. The component with even symmetry is the cosine component, while that with odd symmetry is the sine component.

7.2.2 Source Size and Minimum Spacing

Use Eq. (88) in the following. For an idealized source of shape $I(\nu_0) = I_0$ for $\theta < \theta_0$ and $I(\nu_0) = 0$ for $\theta > \theta_0$, we take the primary beam size of each antenna to be much larger, and define the fringe width for a baseline B θ_b to be $\frac{\lambda}{B}$, The result is

$$R(B) = A\, I_0 \cdot \theta_0 \, \exp\left[\mathrm{i}\,\pi \frac{\theta_0}{\theta_b}\right]\left[\frac{\sin\,(\pi\theta_0/\theta_b)}{(\pi\theta_0/\theta_b)}\right] . \tag{91}$$

The first terms are normalization and phase factors. The important term, in the second set of brackets, is a $\sin x/x$ function. If $\theta_0 >> \theta_b$, the interferometer response is reduced. This is sometimes referred to as the problem of "missing short spacings".

7.2.3 Bandwidth and Beam Narrowing

In Sect. 7.2.1, we noted that on the *white light fringe* the compensation must reach a certain accuracy to prevent a reduction in the interferometer response. However for a finite primary antenna beamwidth, A, this cannot be the case over the entire beam. For two different wavelengths λ_l and λ_s, there will be a phase difference

$$\Delta\phi = 2\pi\, d \left[\frac{\sin\,(\theta_{\text{offset}})}{\lambda_s} - \frac{\sin\,(\theta_{\text{offset}})}{\lambda_l}\right]$$

converting the wavelengths to frequencies, and using $\sin\theta = \theta$, we have

$$\Delta\phi = 2\pi\, \theta_{\text{offset}} \frac{d}{c}\, \Delta\nu \, .$$

With use of the relation $d = \frac{\lambda}{\theta_b}$, we have

$$\Delta\phi = 2\pi \frac{\theta_{\text{offset}}}{\theta_b} \frac{\Delta\nu}{\nu} \, . \tag{92}$$

The effect in Eq. (92) is most important for continuum measurements made with large bandwidths. This effect can be reduced if the cross correlation is carried out using a series of contiguous IF sections. For each of these IF sections, an extra delay is introduced to center the response at the value which is appropriate for that wavelength before correlation.

7.3 *Aperture Synthesis*

Aperture Synthesis is a designation for methods used to derive the intensity distribution $I_\nu(\boldsymbol{s})$ of a part of the radio sky from the measured function $R(\boldsymbol{B})$. To accomplish this we must invert the integral equation (87). This involves Fourier transforms. For even simple images, a large number of computations are needed. Thus Aperture Synthesis and digital computing are intimately connected. In addition, a large number of approximations needs to be applied. We will outline the most important steps of this development without, however, claiming completeness.

For imaging, the relevant relation is:

$$\boxed{I'(x, y) = A(x, y)\, I(x, y) = \int_{-\infty}^{\infty} V(u, v, 0)\, \mathrm{e}^{-\mathrm{i}\, 2\pi(ux+vy)}\, \mathrm{d}u\, \mathrm{d}v} \, , \tag{93}$$

where $V(u, v, 0)$ is the normalized Visibility function, obtained from the correlated and calibrated output of pairs of antennas which are separated by the distances (u, v), expressed in wavelengths. $I'(x, y)$ is the intensity $I(x, y)$ modified by the primary beam shape $A(x, y)$. One can easily correct $I'(x, y)$ by dividing by $A(x, y)$.

Important definitions are:

(1) *Dynamic Range*: The ratio of the maximum to the minimum intensity in an image. In images made with an interferometer array, it should be assumed that the corrections for primary beam taper have been applied. If the minimum intensity is determined by the random noise in an image, the dynamic range is defined by the signal to noise ratio of the maximum feature in the image. The dynamic range is an indication of the ability to recognize low intensity features in the presence of intense features. If the minimum noise is determined by artefacts, i.e., noise in excess of the theoretical noise, the image can be improved by “image improvement techniques”.
(2) *Image Fidelity*: This is defined by the agreement between the measured results and the actual (so-called “true”) source structure. A quantitative comparison would be

$$F = |(S - R)|/S \, ,$$

where F is the fidelity, R is the resulting image obtained from the measurement, and S is the actual source structure. Of course one cannot have a priori knowledge of the correct source structure. In the case of simulations, S is a source model, R is the result of processing S through R.

7.3.1 Interferometric Observations

Usually measurements are carried out in 1 of 3 ways.

- In the first procedure, measurements of the source of interest and a calibrator are made. This is as in the case of single telescope position switching. Two significant differences with single dish measurements are that the interferometer measurement may have to extend over a wide range of hour angles to provide a better coverage of the uv plane, and that instrumental phase must be determined also. One first measures a calibration source or reference source, which has a known position and size, to remove the effect of instrumental phases in the instrument and atmosphere and to calibrate the amplitudes of the sources in question. Sources and calibrators are usually observed alternately. The time variations caused by instrumental and weather effects must be slower than the time between measurements of source and calibrator. If, as is the case for mm/sub-mm wavelength measurements, weather has an important influence, one must switch frequently between target source and calibration source. In *fast switching* one might spend 10 seconds on a nearby calibrator, then a few minutes on-source. This method will reduce the amount of phase fluctuations, but also the amount time available for source measurements. For more rapid changes in the earth's atmosphere, one can correct the phase using measurements of atmospheric water vapor, or changes in the system noise temperature of the individual receivers caused by atmospheric e effects. The corrections for instrumental amplitudes and phases are assumed to be constant over the time when the source is observed. The ratio of amplitudes of source and calibrator are taken to be the normalized source amplitudes. Since the calibrators have known flux densities and positions, the flux densities and positions of the sources can be determined. The reference source should be as close to the on-source as possible, but must have a large enough intensity to guarantee a good signal-to-noise ratio after a short time. Frequently nearby calibrators are time variable over months, so a more distant calibrator with a known or fixed flux density is measured at the beginning or end of the session. This source is usually rather intense, so may also serve as a bandpass calibrator for spectral line measurements. The length of time spent on the off-source measurement is usually no more than few minutes.

- In the second procedure, the so-called *snapshots*, one makes a series of short observations (at different hour angles) of one source after another, and then repeats the measurements. For sensitivity reasons these are usually made in the radio continuum or intense maser lines. As in the first observing method, one intersperses measurements of a calibration source which has a known position and size to remove the effect of instrumental phases in the instrument and atmosphere and to calibrate the amplitudes of the sources in question. The images are affected by the shape of the synthesized beam of the interferometer system. If the size of the source to be imaged is comparable to the primary beam of the individual telescopes, the power pattern of the primary beams will have a large effect. This effect can be corrected easily.
- In the third procedure, one aims to produce a high-resolution image of a source where the goal is either high dynamic range or high sensitivity. The *dynamic range* is the ratio of the highest to the lowest brightness level of reliable detail in the image. This may depend on the signal-to-noise ratio for the data, but for centimeter aperture synthesis observations, spurious features in the image caused by the incomplete sampling of the (u, v) plane are usually more important than the noise. Frequently one measures the source in a number of different interferometer

configurations to better fill the uv plane. These measurements are taken at different times and after calibration, the visibilities are entered into a common data set.

An extension of this procedure may involve the measurement of adjacent regions of the sky. This is *mosaicing*. In a mosaic, the primary beams of the telescopes should overlap, ideally this would be at the half power point. In the simplest case, the images are formed separately and then combined to produce an image of the larger region.

In order to eliminate the loss of source flux density due to missing short spacings, one must supplement the interferometer data with single dish measurements. The diameter of the single dish telescope should be larger than the shortest spacing between interferometer dishes. This single dish image must extend to the FWHP of the smallest of the interferometer antennas. When Fourier transformed and appropriately combined with the interferometer response, this data set has *no* missing flux density. Even in a data set containing single dish data, there are "missing spacings". Improvements that can be applied to images produced from such data sets will be surveyed next.

7.3.2 Real Time Improvements of Visibility Functions

Ideally the relation between the measured $\widetilde{V_{ik}}$ visibility and the *actual* visibility V_{ik} can be considered to be linear

$$\widetilde{V_{ik}}(t) = g_i(t)\, g_k^*(t)\, V_{ik} + \varepsilon_{ik}(t)\ . \tag{94}$$

Average values for the antenna gain factors g_k and the noise term $\varepsilon_{ik}(t)$ are determined by measuring calibration sources as frequently as possible. Actual values for g_k are then computed by linear interpolation. These methods make full use of the fact that the (complex) gain of the array is obtained by multiplication the gains of the individual antennas. If the array consists of n such antennas, $n(n-1)/2$ visibilities can be measured simultaneously, but only $(n-1)$ independent gains g_k are needed (for one antenna, one can arbitrarily set $g_k = 1$ as a reference). In an array with many antennas, the number of antenna pairs greatly exceeds the number of antennas. For phase, one must determine n phases.

Often these conditions can be introduced into the solution in the form of *closure errors*. Introducing the phases φ, θ and ψ by

$$\begin{aligned} \widetilde{V_{ik}} &= |\widetilde{V_{ik}}|\ \exp\{\mathrm{i}\,\varphi_{ik}\}\ , \\ G_{ik} &= |g_i|\,|g_k|\ \exp\{\mathrm{i}\,\theta_i\}\exp\{-\mathrm{i}\,\theta_k\}\ , \\ V_{ik} &= |V_{ik}|\ \exp\{\mathrm{i}\,\psi_{ik}\}\ . \end{aligned} \tag{95}$$

From Eq. (94) the visibility phase ψ_{ik} on the baseline ik will be related to the observed phase φ_{ik} by

$$\varphi_{ik} = \psi_{ik} + \theta_i - \theta_k + \varepsilon_{ik}\ , \tag{96}$$

where ε_{ik} is the phase noise. Then the *closure phase* Ψ_{ikl} around a closed triangle of baseline ik, kl, li,

$$\Psi_{ikl} = \varphi_{ik} + \varphi_{kl} + \varphi_{li} = \psi_{ik} + \psi_{kl} + \psi_{li} + \varepsilon_{ik} + \varepsilon_{kl} + \varepsilon_{li} \ , \tag{97}$$

will be independent of the phase shifts θ introduced by the individual antennas and the time variations. With this procedure, on can minimize phase errors.

Closure amplitudes can also be formed. If four or more antennas are used simultaneously, then ratios, the so-called *closure amplitudes*, can be formed. These are independent of the antenna gain factors:

$$A_{klmn} = \frac{|V_{kl}||V_{mn}|}{|V_{km}||V_{ln}|} = \frac{|\Gamma_{kl}||\Gamma_{mn}|}{|\Gamma_{km}||\Gamma_{ln}|} \ . \tag{98}$$

Both phase and closure amplitudes can be used to improve the quality of the complex visibility function.

If each antenna introduces an unknown complex gain factor g with amplitude and phase, the total number of unknown factors in the array can be reduced significantly by measuring closure phases and amplitudes. If four antennas are available, 50% of the phase information and 33% of the amplitude information can thus be recovered; in a 10 antenna configuration, these ratios are 80 and 78% respectively.

7.3.3 Multi-antenna Array Calibrations

For two antenna interferometers, phase calibration can only be made pair-wise. This is referred to as "baseline based" solutions for the calibration. For a multi-antenna system, there are other and better methods. One can use sets of three antennas to determine the best phase solutions and then combine these to optimize the solution for each antenna. For amplitudes, one can combine sets of four antennas to determine the best amplitude solutions and then optimize this solution to determine the best solution. This process leads to "antenna based" solutions. Antenna based calibrations are used in most cases. These are determined by applying phase and amplitude closure for subsets of antennas and then making the best fit for a given antenna. It is important to note that because of Eq. (50) the output of an antenna can be amplified without seriously degrading the signal-to-noise ratio. For this reason, cross-correlations between a large number of antennas is possible in the radio and mm/sub-mm ranges. This is *not* the case in the optical or near-IR ranges.

7.4 Interferometer Sensitivity

The random noise limit to an interferometer system is calculated following the method used for a single telescope [36]. The RMS fluctuations in antenna temperature are

$$\Delta T_{\mathrm{A}} = \frac{M\, T_{\mathrm{sys}}}{\sqrt{t\, \Delta\nu}}\,, \tag{99}$$

where M is a factor of order unity used to account for extra noise from analog to digital conversions, digital clipping etc. If we next apply the definition of flux density, S_ν, in terms of antenna temperature for a two-element system, we find:

$$\Delta S_\nu = 2\,k\,\frac{T_{\mathrm{sys}}\,\mathrm{e}^{\tau}}{A_{\mathrm{e}}\sqrt{2t\,\Delta\nu}}\,, \tag{100}$$

where τ is the atmospheric optical depth and A_{e} is the effective collecting area of a single telescope of diameter D. There is additional in this expression, since a multiplying interferometer does not process all of the information (i.e., the total power) that the antennas receive. In this case, there is an additional factor of $\sqrt{2}$ compared to the noise in a single dish with an equivalent collecting area, since there is information not collected by a multiplying interferometer. We denote the system noise corrected for atmospheric absorption by $T'_{\mathrm{sys}} = T_{\mathrm{sys}}\,\mathrm{e}^{\tau}$, in order to simplify the following equations. For an array of n identical telescopes, there are $N = n(n-1)/2$ simultaneous pair-wise correlations. Then the RMS variation in flux density is

$$\Delta S_\nu = \frac{2\,M\,k\,T'_{\mathrm{sys}}}{A_{\mathrm{e}}\sqrt{2\,N\,t\,\Delta\nu}}\,. \tag{101}$$

This relation can be recast in the form of brightness temperature fluctuations using the Rayleigh–Jeans relation

$$S = 2\,k\,\frac{T_{\mathrm{b}}\,\Omega_{\mathrm{b}}}{\lambda^2}\,. \tag{102}$$

Then the RMS brightness temperature, due to random noise, in aperture synthesis images is

$$\Delta T_{\mathrm{b}} = \frac{2\,M\,k\,\lambda^2\,T'_{\mathrm{sys}}}{A_{\mathrm{e}}\Omega_{\mathrm{b}}\sqrt{2\,N\,t\,\Delta\nu}}\,. \tag{103}$$

For a Gaussian beam, $\Omega_{\mathrm{MB}} = 1.133\,\theta^2$, so we can relate the RMS temperature fluctuations to observed properties of a synthesis image. The Atacama Large Millimeter/Sub-millimeter Array (ALMA, see Fig. 11) sensitivity calculator is to be found at https://almascience.eso.org/proposing/sensitivity-calculator.

At shorter wavelengths, the RMS temperature fluctuations are lower. Thus, for the same collecting area and system noise, if weather changes are unimportant, a millimeter image should be more sensitive than an image made at centimeter wavelengths. If the effective collecting area remains the same and for a larger main beam solid angle, temperature fluctuations will decrease. For this reason, smoothing an image will result in a lower RMS noise in an image. However, if smoothing is too extreme, this process effectively leads to a decrease in collecting area; then there will be no further improvement in sensitivity.

Fig. 11 This shows seventeen 12-m antennas on the Atacama Large Millimeter/Sub-mm Array (ALMA) site. ALMA is the most ambitious construction project in radio astronomy. ALMA is being built in north Chile on a 5.1 km high site. When finished, ALMA will consist of fifty-four 12-m and twelve 7-m antennas, operating in 10 bands between wavelengths 1 and 0.3 mm. The collecting area of ALMA is 50% that of the Very Large Array of the NRAO, but is vastly more ambitious because of the more complex receivers, the need for highly accurate antennas, and the high altitude site. In addition, the data rates will be orders of magnitude higher than any existing interferometer. At the longest antenna spacing, and shortest wavelength, the angular resolution will be $\approx$5 milli-arcseconds. Photo courtesy of NRAO/ESO/NAOJ

The temperature sensitivity (in Kelvins) for higher angular resolution is worse than for a single telescope with an equal collecting area. From the Rayleigh–Jeans relation, the sensitivity in Jansky (Jy) is fixed by the antenna collecting area and the receiver noise, so only the wavelength and the angular resolution can be varied. Thus, the increase in angular resolution is made at the expense of temperature sensitivity. All other effects being equal, at shorter wavelengths one gains in temperature sensitivity.

Compared to single dishes, interferometers have the great advantage that uncertainties such as pointing and beam size depend fundamentally on timing. Such timing uncertainties can be made very small compared to all other uncertainties. In contrast, the single dish measurements are critically dependent on mechanical deformations of the telescope. In summary, the single dish results are easier to obtain, but source positions and sizes on arcsecond scales are difficult to estimate. The interferometer system has a much greater degree of complexity, but allows one to measure such fine details. The single dish system responds to the source irrespective of the relation of source to beam size; the correlation interferometer will not record source structures larger than a few fringes.

Aperture synthesis is based on sampling the visibility function $V(u, v)$ with separate antennas to provide samples in the (u, v) plane. Many configurations are possible, but the goal is the densest possible coverage of the (u, v) plane. If one calculates the RMS noise in a synthesis image obtained by simply Fourier transforming the (u, v) data, one usually finds a noise level many times higher than that given by Eq. (103) or Eq. (101). There are various reasons for this. One cause is phase fluctuations due to atmospheric or instrumental influences such as LO instabilities. Another cause is due to incomplete sampling of the (u, v) plane. This gives rise to instrumental features, such as stripe-like features in the final images. Yet another systematic effect is the presence of grating rings around more intense sources; these are analogous to *high side lobes* in single dish diffraction patterns. Over the past 20 years, it has been found that these effects can be substantially reduced by software techniques such as CLEAN and Maximum Entropy.

7.4.1 Post Real Time Improvements of Visibility Functions

Before applying specialized techniques, the data must be organized in a useful way without lowering the signal-to-noise ratios. To speed up computations for inverting Eq. (93) one uses the Cooley-Tukey fast Fourier transform algorithm. In order to use the FFT in its simplest version, the visibility function must be placed on a regular grid with sizes that are powers of two of the sampling interval. Since the observed data seldom lie on such regular grids, an interpolation scheme must be used. If the measured points are randomly distributed, this interpolation is best carried out using a convolution procedure.

If some of the spatial frequencies present in the intensity distribution are not present in the (u, v) plane data, then changing the amplitude or phase of the corresponding visibilities will not have any effect on the reconstructed intensity distribution – these have been eliminated. The extent of this effect is shown by the "dirty beam".

Expressed mathematically, if Z is an intensity distribution containing only the unmeasured spatial frequencies, and P_{D} is the dirty beam, then $P_{\mathrm{D}} \otimes Z = 0$.

Hence, if I is a solution of the convolution equation (105) then so is $I + \alpha Z$ where α is any number. This shows that there is no unique solution to the problem.

The solution with visibilities $V = 0$ for the unsampled spatial frequencies is usually called the *principal solution*, and it differs from the true intensity distribution by some unknown *invisible* or *ghost distribution*. It is the aim of image reconstruction to obtain reasonable approximations to these *ghosts* by using additional knowledge or plausible extrapolations, but there is no direct way to select the "best" or "correct" image from all possible images. The familiar linear deconvolution algorithms are not adequate and nonlinear techniques must be used.

The result obtained from the gridded uv data can be Fourier transformed to obtain an image with a resolution corresponding to the size of the array. However, this may still contain artifacts caused by the details of the observing procedure, especially the limited coverage of the (uv) plane. Therefore the dynamic range of such so-called *dirty* maps is rather small. This can be improved by further data analysis, as will be described next.

If the calibrated visibility function $V(u, v)$ is known for the full (u, v) plane both in amplitude and in phase, this can be used to determine the (modified) intensity distribution $I'(x, y)$ by performing the Fourier transformation (93). However, in a realistic situation $V(u, v)$ is only sampled at discrete points within a radius $\cong u_{\max}$ along elliptical tracks, and in some regions of the (u, v) plane, $V(u, v)$ is not measured at all.

We can weight the visibilities by a grading function, g. Then for a discrete number of visibilities, we have a version of Eq. (93) involving a summation, not an integral, to obtain an image via a discrete Fourier transform (DFT):

$$I_{\mathrm{D}}(x, y) = \sum_k g(u_k, v_k) V(u_k, v_k)\, \mathrm{e}^{-\mathrm{i}2\pi(u_k x + v_k y)} , \tag{104}$$

where $g(u, v)$ is a weighting function called the grading or apodisation. To a large extent $g(u, v)$ is arbitrary and can be used to change the effective beam shape and side lobe level. There are two widely used weighting functions: natural and uniform. Natural weighting uses $g(u_k, v_k) = 1$, while Uniform weighting uses $g(u_k, g_k) = 1/N_{\mathrm{s}}(k)$, where $N_{\mathrm{s}}(k)$ is the number of data points within a symmetric region of the (u, v) plane. In a simple case $N_{\mathrm{s}}(k)$ would be a square centered on point k. Data which are naturally weighted result in lower angular resolution but give a better signal-to-noise ratio than uniform weighting. But these are only extreme cases. One can choose intermediate weighting schemes. These are often referred to as *robust* weighting (in the nomenclature of the AIPS data reduction package). Often the reconstructed image I_{D} may not be a particularly good representation of I', but these are related. In another form, Eq. (104) is

$$I_{\mathrm{D}}(x, y) = P_{\mathrm{D}}(x, y) \otimes I'(x, y) , \tag{105}$$

where

$$P_{\mathrm{D}} = \sum_k g(u_k, v_k)\, \mathrm{e}^{-\mathrm{i}2\pi(u_k x + v_k y)} \tag{106}$$

is the response to a point source. This is the *point spread function* PSF for the dirty beam. Thus the *dirty beam* can be understood as a transfer function that distorts the image. (The dirty beam, P_{D}, is produced by the Fourier transform of a point source in the regions sampled; this is the response of the interferometer system to a point source). That is, the *dirty map* $I_{\mathrm{D}}(x, y)$ contains only those spatial frequencies (u_k, v_k) where the visibility function has been measured. The sum in Eq. (106) extends over the same positions (u_k, v_k) as in Eq. (104), and the side lobe structure of the beam depends on the distribution of these points.

7.5 *Advanced Image Improvement Methods*

Digital computing is a crucial part of synthesis array data processing. A large part of the advances in radio synthesis imaging during the last 20 years relies on the progress made in the field of image restoration. In the following we present a few schemes that are applied to improve radio images. However this is by no means an exhaustive collection.

7.5.1 Self-calibration

Amplitude and phase errors scatter power across the image, giving the appearance of enhanced noise. Quite often this problem can be alleviated to an impressive extent by the method of *self-calibration*. This process can be applied if there is a sufficiently intense source in the field contained within the primary beam of the interferometer system. Basically, self-calibration is analogous to using the focus of a camera to sharpen up an object in the field of view. One can restrict the self-calibration to an improvement of phase alone or to both phase and amplitude. Self-calibration is carried in the (u, v) plane. If properly used, this method leads to a great improvement in interferometer images of sources measured with a good signal-to-noise ratio, such as the more intense continuum sources, or of masering spectral lines. If this method is used on objects with low signal-to-noise ratios this method may give very wrong results by concentrating random noise into one part of the interferometer image (see [7]).

In measurements of weak spectral lines, the self-calibration is carried out with a continuum source in the field. The corrections are then applied to the spectral line data. In the case of intense lines, one of the frequency channels containing the emission is used. If self-calibration is applied, the source position information is usually lost.

7.5.2 Applying CLEAN to the Dirty Map

CLEANing is the most commonly used technique to improve single radio interferometer images [20]. This is used after self-calibration or phase/amplitude closure. The *dirty map* is a representation of the principal solution, but with shortcomings. In addition to its inherent low dynamic range, the dirty map often contains features such as negative intensity artifacts. These cannot be real. Another unsatisfactory aspect is that the principal solution is quite often rather unstable, in that it can change drastically when more visibility data are added. Instead of a principle solution that assumes $V = 0$ for all unmeasured visibilities, values for V should be adopted at these positions in the (u, v) plane. These are obtained from some plausible model for the intensity distribution.

The CLEAN method approximates the actual but unknown intensity distribution $I(x, y)$ by the superposition of a finite number of point sources with positive intensity A_i placed at positions (x_i, y_i). The goal of CLEAN is to determine the $A_i(x_i, y_i)$ so that

$$I''(x, y) = \sum_i A_i\, P_{\mathrm{D}}(x - x_i, y - y_i) + I_\varepsilon(x, y)\,, \tag{107}$$

where I'' is the dirty map obtained from the inversion of the visibility function and P_{D} is the dirty beam (Eq. (106)). $I_\varepsilon(x, y)$ is the residual brightness distribution after decomposition. Approximation (107) is deemed successful if I_ε is of the order of the noise in the measured intensities. This decomposition cannot be done analytically, rather an iterative technique has to be used.

The CLEAN algorithm is most commonly applied in the image plane. This is an iterative method which functions in the following fashion: First find the peak intensity of the dirty image, then subtract a fraction γ with the shape of the dirty beam from the image. Then repeat this n times. This *loop gain* $0 < \gamma < 1$ helps the iteration converge, and the iteration process is continued until the intensities of the remaining peaks are below some limit. Usually the resulting point source model is convolved with a *clean beam* of Gaussian shape with a FWHP similar to that of the dirty beam. Whether this algorithm produces a realistic image, and how trustworthy the resulting point source model really is, are unanswered questions.

7.5.3 Maximum Entropy Deconvolution Method (MEM)

The Maximun Entropy Deconvolution Method (MEM) is commonly used to produce a single optimal image from a set of separate but contiguous images [16]. This is used after self-calibration or phase/amplitude closure has been applied to the data. The problem of how to select the "best" image from many possible images which all agree with the measured visibilities is solved by MEM. Using MEM, those values of the interpolated visibilities are selected, so that the resulting image is consistent with all relevant data, including models of the source based on previous results. In addition, the MEM image has maximum smoothness. This is obtained by maximizing the *entropy* of the image. One possible definition of entropy is given by

$$\mathcal{H} = -\sum_i I_i \left[\ln\left(\frac{I_i}{M_i}\right) - 1\right], \tag{108}$$

where I_i is the deconvolved intensity and M_i is a reference image incorporating all "a priori" knowledge. In the simplest case M_i is the empty field $M_i = \mathrm{const} > 0$, or perhaps a lower angular resolution image.

Additional constraints might require that all measured visibilities should be reproduced exactly, but in the presence of noise such constraints are often incompatible with $I_i > 0$ everywhere. Therefore the MEM image is usually constrained to fit the data such that

$$\chi^2 = \sum \frac{|V_i - V_i'|^2}{\sigma_i^2} \tag{109}$$

has the expected value, where V_i is the measured visibility, V_i' is a visibility corresponding to the MEM image and σ_i is the error of the measurement.

8 Continuum Emission from mm/sub-mm Sources

In the early days of mm/sub-mm wavelength astronomy receiver sensitivities restricted measurements to a few continuum sources. This has improved dramatically with the use of semiconductor bolometers pioneered by F. J. Low. Subsequently spectral lines of molecules such as CO, HCN and CS were found. These were rather intense, and this led to a blossoming of the field. Subsequently, spectral lines of neutral carbon and a line of ionized carbon were detected. We sketch continuum radiation mechanisms and then spectral line mechanisms that are specific to the mm/sub-mm wavelength range.

Radio sources can thus be classified into two categories: those which radiate by thermal mechanisms and the others, which radiate by nonthermal processes. In practice, non-thermal radiation can be explained by the synchrotron process. This is caused by relativistic electrons spiraling in a magnetic field (see [41] for a delightful and instructive account). For optically thin synchrotron radiation, most sources have a flux density that varies as

$$\boxed{S_\nu = S_0 \left(\frac{\nu}{\nu_0}\right)^{-\alpha}}, \tag{110}$$

where α has a positive value. Synchrotron radiation often shows linear polarization, and in rare cases, shows circular polarization. The most prominent example of a synchrotron source with $\alpha = 0$, i.e., a flat spectrum is Sgr A*. This source, at 8.5 kpc from the Sun, is considered to be the closest supermassive Black Hole. The variation of flux density with frequency may be an indication that the synchrotron emission is optically thick, and/or has a non-standard geometry (see e.g. [29]). Interferometry at 1.3 mm of Sgr A* with baselines of up to a few 10^3 km is ongoing. Measurements of synchrotron emission give only a very limited set of source parameters. When combined with other data, synchrotron measurements can be used to obtain source parameters (see [52]).

The most famous example of black body radiation is the 2.73 K cosmic microwave background, CMB. This source of radiation is fit by a Planck curve to better than 0.1%. The difficulty in detecting it was *not* due to the weakness of the signal, but rather due to the fact that the radiation is present in all directions, so that scanning a telescope over the sky and taking differences will not lead to a detection. The actual discovery was obtained by a comparison of measurements of the noise temperature from the sky, the receiver and the ground with the absolute temperature of a helium cooled load, using Dicke switching. Studies of the CMB are conducted from satellites and balloons; these are directed at determinations of the polarization and deviations from the Black Body spectrum.

As an example of thermal radiation, we consider planets. These are black bodies with spectra that are an almost exact representation of the Rayleigh–Jeans law for various temperatures. For H II regions such as Orion A (M 42, NGC 1976) the spectrum is not a simple black body, but the explanation is fairly straightforward. If we consider the solution of the equation of radiation transfer (14) for an isothermal object without a background source

$$I_\nu = B_\nu(T)\,(1 - \mathrm{e}^{-\tau_\nu})\,,$$

we find that $I_\nu < B_\nu$ if $\tau_\nu < 1$; the frequency variation of τ_ν is

$$\boxed{\tau_\nu = 3.014 \times 10^{-2} \left(\frac{T_\mathrm{e}}{\mathrm{K}}\right)^{-3/2} \left(\frac{\nu}{\mathrm{GHz}}\right)^{-2} \left(\frac{\mathrm{EM}}{\mathrm{pc\ cm^{-6}}}\right) \langle g_\mathrm{ff}\rangle}\,, \tag{111}$$

where T_e is the electron temperature and the Gaunt factor for free-free transitions is given by

$$\boxed{\langle g_\mathrm{ff}\rangle = \begin{cases} \ln\left[4.955 \times 10^{-2} \left(\frac{\nu}{\mathrm{GHz}}\right)^{-1}\right] + 1.5 \ln\left(\frac{T_\mathrm{e}}{\mathrm{K}}\right) \\ 1 \quad \text{for} \quad \frac{\nu}{\mathrm{MHz}} \gg \left(\frac{T_\mathrm{e}}{\mathrm{K}}\right)^{3/2} \end{cases}}\,. \tag{112}$$

The term EM is

$$\mathrm{EM} = \int N_e^2\,\mathrm{d}l$$

with units cm^{-6} pc. This is a complex relation and was derived only after plowing through lots and lots of math. N_e cannot be directly obtained from EM and the source size because of clumping.

In the mm/sub-mm, $\tau_\nu \ll 1$ is usually the case, that is, optically thin. Then, converting from I_ν to T_b, we have:

$$\boxed{T_\mathrm{b} = T_\mathrm{e}\,\tau_\nu = 3.014 \times 10^{-2} \left(\frac{T_\mathrm{e}}{\mathrm{K}}\right)^{-1/2} \left(\frac{\nu}{\mathrm{GHz}}\right)^{-2} \left(\frac{\mathrm{EM}}{\mathrm{pc\ cm^{-6}}}\right) \langle g_\mathrm{ff}\rangle}\,. \tag{113}$$

We next turn our attention to emission from dust grains. It appears that cold dust, with temperatures 10–30 K, makes up much of the mass of dust and, by implication, traces cold interstellar gas. The gas mass can be estimated, given a dust-to-gas ratio. Since the average size of a grain is thought to be ~0.3 μm, the wavelength of the radiation is much longer than the size of the emitter. For this reason, the efficiency of sub-mm radiation is rather low. Radiation from dust grains shows linear polarization, which leads to the conclusion that grains are elongated and are aligned by magnetic fields [26].

If we use the exact relation, we have

$$T_b(\nu) = T_0 \left(\frac{1}{e^{T_0/T_{dust}} - 1} - \frac{1}{e^{T_0/2.7} - 1} \right) (1 - e^{-\tau_{dust}}) , \qquad (114)$$

where $T_0 = h\nu/k$. This is completely general; if we neglect the 2.7 K background,

$$T_{dust} = T_0 \left(\frac{1}{e^{T_0/T_{dust}} - 1} \right) (1 - e^{-\tau_{dust}}) . \qquad (115)$$

If $T_{dust} \gg T_0$, we can simplify this expression, but in the mm/sub-mm range this may *not* be the case. However, the most important step is in making a quantitative connection between τ_{dust} and N_{H_2}. The relation between τ_{dust} and the gas column density must be determined empirically. Unlike the planets, which have measured sizes, the radiation from dust grains depends on the surface area of the grains, which cannot be determined directly. If a relation between dust mass and τ can be determined, it is simple to convert to the total mass, since dust is generally accepted to be between 1/100 and 1/150th of the total mass. All astronomical determinations are based on [19]. An updated version of the relation given by [31] is:

$$\tau_{dust} = 8.5 \times 10^{-21} \frac{Z}{Z_\odot} b\, N_H\, \lambda^{-\beta} , \qquad (116)$$

where λ is the wavelength in μm and is >100 mm, N_H is in units cm^{-2}, and Z is the metallicity as a ratio of that of the sun $Z_\odot$. In the cm and mm wavelength range, the dust optical depth is small with β values between 1 and 2; β is often taken to be 2, but this assumption is to be viewed with caution. The parameter b is an adjustable factor used to take into account changes in grain sizes. Currently, it is believed that $b = 1.9$ is appropriate for moderate density gas and $b = 3.4$ for dense gas (but this is not certain). At long millimeter wavelengths, a number of observations have shown that the optical depth of such radiation is small. Then the observed temperature is

$$T = T_{dust}\, \tau_{dust} , \qquad (117)$$

where the quantities on the right side are the dust temperature and optical depth. Then the flux density is

$$S = \frac{2\,k\,T}{\lambda^2} = 2\,k\,T_{\mathrm{dust}}\lambda^{-2}\,\tau_{\mathrm{dust}}\,\Delta\Omega\ . \tag{118}$$

If the dust radiation is expressed in mJy, the source FWHP size, θ, in arcseconds, and the wavelength, λ, in mm, the column density of hydrogen in all forms, N_{H}, in the Rayleigh–Jeans approximation is:

$$N_{\mathrm{H}} = 1.55 \times 10^{24}\,\frac{S_\nu}{\theta^2}\,\frac{\lambda^4}{Z/Z_\odot\,b\,T_{\mathrm{dust}}}\ . \tag{119}$$

This relation is based on the assumption that the emissivity increases as $\lambda^{-\beta}$ with a β value of 2; in some cases β may be in the range 1 to 1.5. In any case, the flux density increases rapidly at shorter wavelengths, dust emission will become more important at millimeter and sub-mm wavelengths. An illustration of broadband dust emission from redshifted sources is shown in Fig. 12.

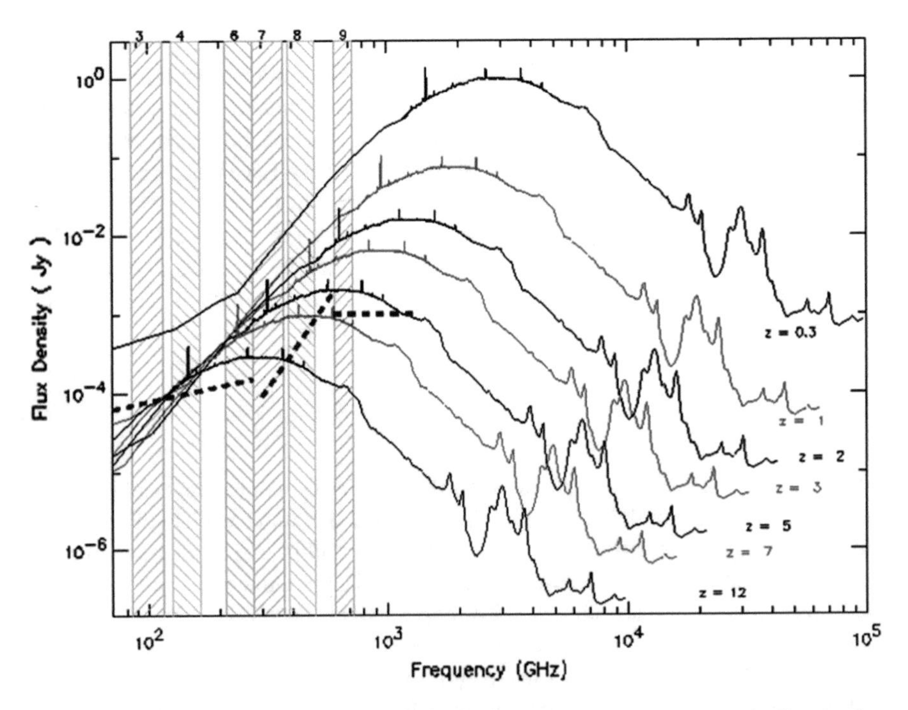

Fig. 12 The spectrum of the starburst galaxy M82. This galaxy is rather close to the Sun, but has a star formation rate that is much higher than for the Milky Way. The large featureless maximum is caused by dust radiation. On top of this are various molecular and atomic lines. The different curves represent M82 for the labelled redshifts. Although the emission decreases inversely with the square of the distance, this is compensated by the shift of dust radiation to lower frequencies. These effects allow one to detect an oject such as M82 to redshift $z = 12$ with the Atacama Large Millimeter Array (ALMA; see also Fig. 11). The ALMA receiver bands are indicated by numbers in the upper left side of this figure (P. Cox, unpublished)

In summary, dust grains emit radiation following a modified Planck law. The intensity, expressed as flux density per beam, varies as $\lambda^{-2-\beta}$, where β is between 1 and 2. Thus, the intensity increases rapidly in the mm and sub-mm ranges. Grains has an average size of $\sim$0.3 μm, so in the mm/sub-mm or near IR range the wavelength is much larger than the grain size. Thus the radiation process is rather inefficient. Complicating an analysis of the spectrum is also the fact that for a given source there are usually a range of temperatures. Dust emission has been found to be polarized. From this, it was concluded that the grains are non-spherical and aligned by an interstellar magnetic field. It is thought that the long axis of the grains is approximately perpendicular to the direction of the field, so the direction of B is at right angles to that of the measured broadband linearly polarized radiation emitted by dust. The direction but not the strength of the magnetic field can be determined from measurements of dust polarization. The fractional polarization is rather small, so measurements require high sensitivity and the supression of instrumental effects. Recently, very weak excess radiation from dust at frequencies between about 10 to 50 GHz, with a peak at about 30 GHz has been reported. This has been identified with emission from Very Small Grains (VSG). See [40] and references therein for an account of these measurements and comparisons with theory.

9 Spectral Line Basics

In local thermodynamic equilibrium (LTE) the intensities of emitted and absorbed radiation are not independent but are related by Kirchhoff's law (9). This applies to both continuous radiation and line radiation. The Einstein coefficients give a convenient means to describe the interaction of radiation with matter by the emission and absorption of photons [38]. These are:

$$\boxed{g_1\,B_{12} = g_2\,B_{21}} \tag{120}$$

and

$$\boxed{A_{21} = \frac{8\pi h\nu_0^3}{c^3}\,B_{21}}\,. \tag{121}$$

9.1 *Radiative Transfer with Einstein Coefficients*

When the radiative transfer was considered in Sect. 2.1, the material properties were expressed as the emission coefficient ε_ν and the absorption coefficient κ_ν. Both ε_ν and κ_ν are macroscopic parameters; for a physical theory these must to be related to atomic properties of the matter in the cavity. If line radiation is considered, the

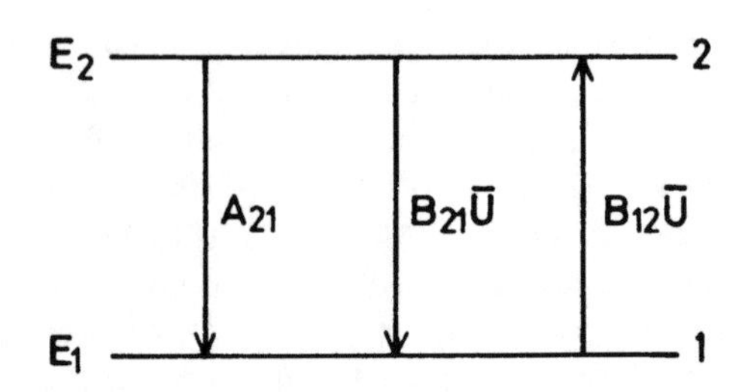

Fig. 13 Transitions between the states 1 and 2 and the Einstein probabilities

Einstein coefficients are very useful because these can be linked directly to the properties of the transition responsible for the spectral line. We show a sketch of a two level system in Fig. 13. For radiative transfer ε_ν and κ_ν are needed, so we must investigate the relation between κ_ν and A_{ik} and B_{ik}. This is best done by considering the possible change of intensity I_ν passing through a slab of material with thickness $\mathrm{d}s$ as in Sect. 2.1. Now we will use A_{ik} and B_{ik}.

According to Einstein there are three different processes contributing to the intensity I_ν. Each system making a transition from E_2 to E_1 contributes the energy $h\nu_0$ distributed over the full solid angle 4π. Then the total amount of energy emitted spontaneously is

$$\mathrm{d}E_\mathrm{e}(\nu) = h\nu_0\, N_2\, A_{21}\, \varphi_\mathrm{e}(\nu)\, \mathrm{d}V \frac{\mathrm{d}\Omega}{4\pi}\, \mathrm{d}\nu\, \mathrm{d}t\ . \tag{122}$$

For the total energy *absorbed* we similarly obtain

$$\mathrm{d}E_\mathrm{a}(\nu) = h\nu_0\, N_1\, B_{12}\, \frac{4\pi}{c}\, I_\nu\, \varphi_\mathrm{a}(\nu)\, \mathrm{d}V \frac{\mathrm{d}\Omega}{4\pi}\, \mathrm{d}\nu\, \mathrm{d}t \tag{123}$$

and for the *stimulated* emission

$$\mathrm{d}E_\mathrm{s}(\nu) = h\nu_0\, N_2\, B_{21}\, \frac{4\pi}{c}\, I_\nu\, \varphi_\mathrm{e}(\nu)\, \mathrm{d}V \frac{\mathrm{d}\Omega}{4\pi}\, \mathrm{d}\nu\, \mathrm{d}t\ . \tag{124}$$

The line profiles $\varphi_\mathrm{a}(\nu)$ and $\varphi_\mathrm{e}(\nu)$ for absorbed and emitted radiation could be different, but in astrophysics it is usually permissible to put $\varphi_\mathrm{a}(\nu) = \varphi_\mathrm{e}(\nu) = \varphi(\nu)$. For the volume element we put $\mathrm{d}V = \mathrm{d}\sigma\, \mathrm{d}s$, where $\mathrm{d}\sigma$ is the unit area perpendicular to the beam direction. For a stationary situation, we find

$$\begin{aligned} &\mathrm{d}E_\mathrm{e}(\nu) + \mathrm{d}E_\mathrm{s}(\nu) - \mathrm{d}E_\mathrm{a}(\nu) = \mathrm{d}I_\nu\, \mathrm{d}\Omega\, \mathrm{d}\sigma\, \mathrm{d}\nu\, \mathrm{d}t \\ &= \frac{h\nu_0}{4\pi}\left[N_2\, A_{21} + N_2\, B_{21}\, \frac{4\pi}{c}\, I_\nu - N_1\, B_{12}\, \frac{4\pi}{c}\, I_\nu\right] \varphi(\nu)\, \mathrm{d}\Omega\, \mathrm{d}\sigma\, \mathrm{d}s\, \mathrm{d}\nu\, \mathrm{d}t\ . \end{aligned}$$

The resulting equation of transfer with Einstein coefficients is

$$\boxed{\frac{\mathrm{d}I_\nu}{\mathrm{d}s} = -\frac{h\nu_0}{c}\,(N_1\, B_{12} - N_2\, B_{21})\, I_\nu\, \varphi(\nu) + \frac{h\nu_0}{4\pi} N_2\, A_{21}\, \varphi(\nu)}\ . \tag{125}$$

Comparing this with Eq. (8) we obtain agreement by putting

$$\boxed{\kappa_\nu = \frac{h\nu_0}{c}\, N_1\, B_{12}\left(1 - \frac{g_1 N_2}{g_2 N_1}\right)\varphi(\nu)} \tag{126}$$

and

$$\boxed{\varepsilon_\nu = \frac{h\nu_0}{4\pi}\, N_2\, A_{21}\,\varphi(\nu)} \,. \tag{127}$$

The factor in brackets in Eq. (126) is the correction for stimulated emission. In radio astronomy, where the stimulated emission almost completely cancels the effect of the true absorption, this is important. How this comes about is best seen if we investigate what becomes of Eqs. (125)–(127) if LTE is assumed.

From Eqs. (126) and (127) we find that

$$\frac{\varepsilon_\nu}{\kappa_\nu} = \frac{2h\nu^3}{c^2}\left(\frac{g_2 N_1}{g_1 N_2} - 1\right)^{-1} .$$

But for LTE, according to Eq. (9), this should be equal to the Planck function, resulting in

$$\boxed{\frac{N_2}{N_1} = \frac{g_2}{g_1}\exp\left(-\frac{h\nu_0}{k\,T}\right)} \,. \tag{128}$$

In LTE, the energy levels are populated according to the same Boltzmann distribution for the temperature T that applied to full TE. Then the absorption coefficient becomes

$$\boxed{\kappa_\nu = \frac{c^2}{8\pi}\frac{1}{\nu_0^2}\frac{g_2}{g_1}\, N_1\, A_{21}\left[1 - \exp\left(-\frac{h\nu_0}{k\,T}\right)\right]\varphi(\nu)} \,, \tag{129}$$

where we have replaced the B coefficient by the A coefficient, using

$$B_{12} = \frac{g_2}{g_1}\, A_{21}\, \frac{c^3}{8\pi h\nu^3} \,.$$

9.2 *Dipole Transition Probabilities*

The simplest sources for electromagnetic radiation are oscillating dipoles. Radiating electric dipoles have already been treated classically, but it should also be possible to express these results in terms of the Einstein coefficients. There are two types of dipoles that can be treated by quite similar means: the electric and the magnetic dipole.

Electric Dipole. Consider an oscillating electric dipole

$$d(t) = e\,x(t) = e\,x_0 \cos\omega t\ . \tag{130}$$

According to electromagnetic theory, this will radiate. The power emitted into a full 4π steradian is

$$P(t) = \frac{2}{3}\frac{e^2\dot{v}(t)^2}{c^3}\ . \tag{131}$$

Expressing $x = d/e$ and $\dot{v} = \ddot{x}$, we obtain an average power, emitted over one period of oscillation of

$$\langle P \rangle = \frac{64\pi^4}{3\,c^3}\,\nu_{mn}^4 \left(\frac{e\,x_0}{2}\right)^2\ . \tag{132}$$

This mean emitted power can also be expressed in terms of the Einstein A coefficient:

$$\langle P \rangle = h\,\nu_{mn}\,A_{mn}\ . \tag{133}$$

Equating (132) and (133) we obtain

$$\boxed{A_{mn} = \frac{64\pi^4}{3\,h\,c^3}\,\nu_{mn}^3\,|\mu_{mn}|^2}\ . \tag{134}$$

This was also cited in Sect. 2.3.1, where

$$\mu_{mn} = \frac{e\,x_0}{2} \tag{135}$$

is the mean electric dipole moment of the oscillator for this transition.

Expression (134) is applicable only to classical electric dipole oscillators, but is also valid for quantum systems.

9.3 *Simple Solutions of the Rate Equation*

In order to compute absorption or emission coefficients in Eqs. (126) and (127), both the Einstein coefficients and the number densities N_i and N_k must be known. In the case of LTE, the ratio of N_2 to N_1 is given by the Boltzmann function:

$$\frac{N_2}{N_1} = \frac{g_2}{g_1}\exp\left(-\frac{h\,\nu_0}{k\,T_\mathrm{K}}\right)\ . \tag{136}$$

If C_{12} and C_{21} are the collision probabilities per particle (in $\mathrm{cm}^3\ \mathrm{s}^{-1}$) for the transitions $1 \to 2$ and $2 \to 1$, respectively, T_{K} is the *kinetic temperature*. When T_{ex}, T_{b} and $T_{\mathrm{K}} \gg h\nu/k$, and if we use for abbreviation

$$T_0 = \frac{h\nu}{k}\,, \tag{137}$$

we have

$$\boxed{T_{\mathrm{ex}} = T_{\mathrm{K}} \frac{T_{\mathrm{b}} A_{21} + T_0 C_{21}}{T_{\mathrm{K}} A_{21} + T_0 C_{21}}}\;. \tag{138}$$

If radiation dominates the rate equation ($C_{21} \ll A_{21}$), then $T_{\mathrm{ex}} \to T_{\mathrm{b}}$. If on the other hand collisions dominate ($C_{21} \gg A_{21}$), then $T_{\mathrm{ex}} \to T_{\mathrm{K}}$. Since C_{ik} increases with N, collisions will dominate the distribution in high-density situations and the excitation temperature of the line will be equal to the kinetic temperature. In low-density situations $T_{\mathrm{ex}} \to T_{\mathrm{b}}$. The density when $A_{21} \approx C_{21} \approx N^*\langle\sigma\, v\rangle$ is called the *critical density*. The smaller A_{21}, the lower is N^*.

Radiative transfer with high optical depths can be dealt with using the Large Velocity Gradient (LVG) approximation. The LVG approximation is the simplest model for such transport. In this, it is assumed that the spherically symmetric cloud possesses large scale systematic motions so that the velocity is a function of distance from the center of the cloud, that is, $V = V_0(r/r_0)$. Furthermore, the systematic velocity is much larger than the thermal line width. Then the photons emitted by a two level system at one position in the cloud can only interact with those that are nearby. Then the global problem of photon transport is reduced to a local problem. With LVG, one has a simple method to estimate the effects of photon trapping. If we neglect the 2.7 K background and use the relation $T_0 = h\,\nu/k$, we have

$$\frac{T}{T_0} = \frac{T_{\mathrm{k}}/T_0}{1 + T_{\mathrm{k}}/T_0\ \ln\left[1 + \dfrac{A_{ji}}{3C_{ji}\,\tau_{ij}}\left(1 - \exp\left(-3\,\tau_{ij}\right)\right)\right]}\,. \tag{139}$$

The term $(1 - \exp(-3\,\tau_{ij}))/\tau_{ij}$ is caused by 'photon trapping' in the cloud. If $\tau_{ij} \gg 1$, the case of interest, then A_{ij} is replaced by A_{ij}/τ_{ij}. All of this information is basic, but does not include the details for specific species, such as energy-level structures, collisional cross sections and A coefficients.

10 Line Radiation from Atoms

Most atomic transitions give rise to spectral lines at infrared or shorter wavelengths. With the exception of radio recombination lines, atomic radio lines are rare. The energy levels are described by the scheme ${}^{2S+1}L_J$. In this description, S is the total

Table 1 Parameters of some atomic lines

Element and ionization state	Transition	ν/GHz	A_{ij}/s^{-1}	Critical density n^*	Notes
CI	${}^3P_1 - {}^3P_0$	492.16	7.93×10^{-8}	5×10^2	b
CI	${}^3P_2 - {}^3P_1$	809.34	2.65×10^{-7}	10^4	b
CII	${}^2P_{3/2} - {}^2P_{1/2}$	1900.54	2.4×10^{-6}	5×10^3	b
OI	${}^3P_0 - {}^3P_1$	2060.07	1.7×10^{-5}	$\sim 4 \times 10^5$	b
OI	${}^3P_1 - {}^3P_2$	4744.77	8.95×10^{-5}	$\sim 3 \times 10^6$	a,b
OIII	${}^3P_1 - {}^3P_0$	3392.66	2.6×10^{-5}	$\sim 5 \times 10^2$	a
OIII	${}^3P_2 - {}^3P_1$	5785.82	9.8×10^{-5}	$\sim 4 \times 10^3$	a
NII	${}^3P_1 - {}^3P_0$	1473.2	2.1×10^{-6}	$\sim 5 \times 10^1$	a
NII	${}^3P_2 - {}^3P_1$	2459.4	7.5×10^{-6}	$\sim 3 \times 10^2$	a
NIII	${}^2P_{3/2} - {}^2P_{1/2}$	5230.43	4.8×10^{-5}	$\sim 3 \times 10^3$	a,b

[a]Ions or electrons as collision partners
[b]H_2 as a collision partner

spin quantum number, and $2S + 1$ is the multiplicity of the line, that is the number of possible spin states. L is the total orbital angular momentum of the system in question, and J is the total angular momentum. For the lighter elements, the energy levels are best described using LS coupling. This is constructed by vectorially summing the orbital momenta to obtain the total $\boldsymbol{L}$, then combining the spins of the individual electrons to obtain $\boldsymbol{S}$, and then vectorially combining $\boldsymbol{L}$ and $\boldsymbol{S}$ to obtain $\boldsymbol{J}$. If the nucleus has a total spin, $\boldsymbol{I}$, this can be vectorially combined with $\boldsymbol{J}$ to form $\boldsymbol{F}$. For an isolated system, all of these quantum numbers have a constant magnitude and also a constant projection in one direction. Usually the direction is arbitrarily chosen to be along the z axis, and the projected quantum numbers are referred to as M_F, M_J, M_L and M_S.

We give a list of the quantum assignments together with line frequencies, Einstein A coefficients and critical densities in Table 1 of millimeter and sub-millimeter atomic lines [36].

The most studied mm/sub-mm atomic lines are those of neutral carbon at 492 and 809 GHz. These lines arise from molecular regions that are somewhat protected from the interstellar ultraviolet radiation. In less obscured regions, ionized carbon, C^+ or C II is present. This ion has a fine structure line at 157 μm and is expected to be a dominant cooling line in denser clouds. Although these lines might be considered as part of infrared astronomy, the heterodyne techniques have reached the 1.4 THz range (= 200 μm) from the ground and 150 μm with the Herschel satellite. The oxygen lines must be measured from high flying aircraft such as SOFIA or satellites.

11 Emission Nebulae, Radio Recombination Lines

The physical state of the interstellar medium varies greatly from one region to the next, because the gas temperature depends on the local energy input. There exist large, cool cloud complexes in which both dust grains and many different molecular species are abundant. Often new stars are born in these dense clouds, and since they are sources of thermal energy, the stars will heat the gas surrounding them. If the stellar surface temperature is sufficiently high, most of the energy will be emitted as photons with $\lambda < 912$ Å. This radiation has sufficient energy to ionize hydrogen. Thus young, luminous stars embedded in gas clouds will be surrounded by emission regions in which the gas temperature and consequently the pressure will be much higher than in cooler clouds. Occasionally an ion will recombine with a free electron. Since the ionization rate is rather low, the time interval between two subsequent ionizations of the same atom will generally be much longer than the time for the electron to cascade to the ground state, and the cascading atom will emit recombination lines.

11.1 Rydberg Atoms

The behavior of Rydberg atoms in the interstellar medium can show complex excitation properties; such systems give an indication of the excitation effects one often finds with molecules [15]. When ionized hydrogen recombines at some level with the principal quantum number $n > 1$, the atom will emit recombination line emission on cascading down to the ground state. The radius of the nth Bohr orbit is

$$a_n = \frac{\hbar^2}{Z^2\, m\, e^2}\, n^2 \,, \tag{140}$$

and so for large principle quantum number n, the effective radius of the atom becomes exceedingly large. Systems in such states are generally called Rydberg atoms. Energy levels in these are quite closely spaced, and since pressure effects at large n caused by atomic collision may become important, the different lines eventually will merge.

The frequency of the atomic lines of hydrogen-like atoms are given by the Rydberg formula

$$\nu_{ki} = Z^2 R_M \left(\frac{1}{i^2} - \frac{1}{k^2}\right) \,, \quad i < k \,, \tag{141}$$

where

$$R_M = \frac{R_\infty}{1 + \dfrac{m}{M}} \tag{142}$$

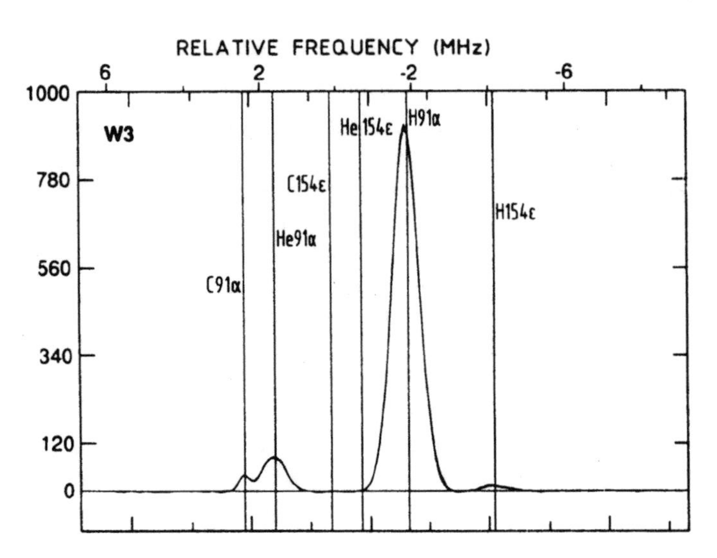

Fig. 14 Recombination lines in the H II region W 3 at 8.5 GHz. The most intense lines are H 91 α, He 91 α, C 91 α, the total integration time, t, for this spectrum is 75 hours; even after this time, the RMS noise follows a theoretical dependence of $1/\sqrt{t}$. Figure taken from Balser et al. (1994) [3]

if m is the mass of the electron, M that of the nucleus and Z is the effective charge of the nucleus in units of the proton charge. For $n > 100$ we always have $Z \approx 1$ and the spectra of all atoms are quite hydrogen-like, the only difference being a slightly changed value of the Rydberg constant.

Lines corresponding to the transitions $n + 1 \rightarrow n$ are most intense and are called α lines. Those for transitions $n + 2 \rightarrow n$ are β lines; $n + 3 \rightarrow n$ transitions are γ lines; etc. In the identification of a line both the element and the principle quantum number of the lower state are given: so H 91 α is the line corresponding to the transition $92 \rightarrow 91$ of H, while H 154 ε corresponds to $158 \rightarrow 154$ of H (see Fig. 14).

All atoms with a single electron in a highly excited state are hydrogen-like. The radiative properties of these Rydberg atoms differ only by their different nuclear masses. The Einstein coefficients A_{ik}, the statistical weights g_i, and the departure coefficients b_i are identical for all Rydberg atoms, if the electrons in the inner atomic shells are not involved. The frequencies of the recombination lines are slightly shifted by the reduced mass of the atom. If this frequency difference is expressed in terms of radial velocities this difference is independent of the quantum number for a given element.

The line width of interstellar radio recombination lines is governed by external effects; neither the intrinsic line width, nor the fine structure of the atomic levels has observable consequences. In normal H II regions, evidence for broadening of the lines by inelastic collisions is found for $N \geq 130$, from the broad line wings. For $N < 60$ the observed linewidth is fully explainable by Doppler broadening. The purely thermal Doppler broadening for hydrogen is:

$$\Delta V_{\frac{1}{2}} = 0.21\sqrt{T_{\mathrm{K}}}\ .$$

The electrons have a velocity distribution that is described very closely by a Maxwellian velocity distribution; long range Coulomb forces eliminate any deviations with a relaxation time that is exceedingly short. This distribution is characterized by an electron temperature T_{e} and, due to the electrostatic forces, the protons should have a similar distribution with the same temperature. The spectral lines are observed to have Gaussian shapes. Thermal Doppler motions for $T_{\mathrm{e}} \cong 10^4$ K produce a line width of 21.4 km s^{-1}, however a width of $\sim$25 km s^{-1}–$\sim$30 km s^{-1} is observed. Therefore it is likely that nonthermal motions in the gas contribute to the broadening. These motions are usually referred to as *micro turbulence*. If we include this effect, the half width of hydrogen is generalized to

$$\Delta V_{\frac{1}{2}} = \sqrt{0.04576\, T_{\mathrm{e}} + v_{\mathrm{t}}^2}\ . \tag{143}$$

11.2 LTE Line Intensities

From the correspondence principle, the data for high quantum numbers can be computed by using classical methods, and therefore we will use for A_{ki} the expression (134) for the electric dipole. For the dipole moment in the transition $n+1 \rightarrow n$, we put

$$\mu_{n+1,n} = \frac{e\, a_n}{2} = \frac{h^2}{8\pi^2 m\, e}\, n^2\ ,$$

where $a_n = a_0\, n^2$ is the Bohr radius of hydrogen, and correspondingly

$$\nu_{n+1,n} = \frac{m\, e^4}{4\pi^2\, h^3\, n^3}\ .$$

Substituting this expression into Eq. (134) we obtain, for the limit of large n

$$A_{n+1,n} = \frac{64\pi^6\, m\, e^{10}}{3\, h^6\, c^3}\frac{1}{n^5} = \frac{5.36\times 10^9}{n^5}\ \mathrm{s}^{-1}\ . \tag{144}$$

We adopt a Gaussian line shape. Introducing the full line width $\Delta\nu$ at half intensity points, we obtain for the optical depth at the line center for an α line

$$\text{EM} = \int N_e(s)\, N_p(s)\, \mathrm{d}s = \int \left(\frac{N_e(s)}{\text{cm}^{-3}}\right)^2 \mathrm{d}\left(\frac{s}{\text{pc}}\right) \tag{145}$$

$$\tau_L = 1.92 \times 10^3 \left(\frac{T_e}{\text{K}}\right)^{-5/2} \left(\frac{\text{EM}}{\text{cm}^{-6}\,\text{pc}}\right) \left(\frac{\Delta\nu}{\text{kHz}}\right)^{-1}, \tag{146}$$

where the emission measure, EM, was calculated using $N_p(s) \approx N_e(s)$. This should be reasonable since the abundance of He is 0.1 that of H. We assume that $\tau_L \ll 1$, and therefore that $T_L = T_e\, \tau_L$, or

$$T_L = 1.92 \times 10^3 \left(\frac{T_e}{\text{K}}\right)^{-3/2} \left(\frac{\text{EM}}{\text{cm}^{-6}\,\text{pc}}\right) \left(\frac{\Delta\nu}{\text{kHz}}\right)^{-1}. \tag{147}$$

For $\tau_c < 1$ (see Eq. (111)) for the continuum, we obtain

$$\frac{T_L}{T_c}\left(\frac{\Delta v}{\text{km s}^{-1}}\right) = \frac{1.91 \times 10^4}{< g_{ff} >} \left[\frac{\nu}{\text{GHz}}\right] \left[\frac{T_e}{\text{K}}\right]^{-1} \frac{1}{1 + N(\text{He}^+)/N(\text{H}^+)}. \tag{148}$$

In this expression, $< g_{ff} >$ is the Gaunt factor (Eq. (112)) and T_e is the LTE electron temperature, denoted as T_e^*. The last factor is due to the fact that both N_{H^+} and N_{He^+} contribute to $N_e = N_{H^+} + N_{He^+}$. Typical values for the H II region Orion A at 100 GHz are $N(H_e^+)/N(H^+) = 0.08$, $T_L/T_c \approx 1$ and $\Delta V_{\frac{1}{2}} = 25.7$ km s^{-1}, which give a T_e^* value of 8200 K. Equation (148) is valid only if both the line and continuum radiation are optically thin. This is the case for nearly all sources in the mm/sub-mm range. One possible exception is the extraordinary case of MWC 349 [30] which shows time-variability and strong maser action in the mm/sub-mm range.

11.3 Non LTE Line Intensities

The *departure coefficients*, b_n, relate the true population of level, N_n, to the population under LTE conditions, N_n^*, by

$$N_n = b_n\, N_n^*. \tag{149}$$

For hydrogen and helium, the b_n factors are always $<$1, since the A coefficient for the lower state is larger and the atom is smaller so collisions are less effective. For states i and k, with $k > i$ we have $b_n \rightarrow 1$ for LTE. For any pair of energy levels, the upper level is always overpopulated relative to the lower level. Since $h\nu \ll kT$ this overpopulation leads to a *negative* excitation temperature. This gives rise to recombination line masering. Usually the line optical depth is very small, but the background continuum could be amplified. In H II regions, the dominant effect is often a slightly lower line intensity which leads to a slight overestimate of the electron temperature of the H II region. If dust emission contributes to the continuum emission, it may be possible that masering of radio recombination lines becomes more prominent.

12 Overview of Molecular Basics

We present the basic concepts needed to understand the radiation from molecules that are widespread in the Interstellar Medium (ISM) [25, 48]. For linear molecules, examples are carbon monoxide, CO, SiO, and N_2H^+; for symmetric top molecules, these are ammonia, NH_3, CH_3CN and CH_3CCH; and for asymmetric top molecules these are water vapor, H_2O, formaldehyde, H_2CO and H_2D^+. Finally we present a short account of molecules with non-zero electronic angular momentum in the ground state, using OH as an example, and then present an account of methanol, CH_3OH, which has hindered motion. In each section, we give relations between molecular energy levels, column densities and local densities. Although molecular line emission is complex, such measurements allow a determination of parameters in heavily obscured regions not accessible in the near-infrared or optical.

12.1 Basic Concepts

The structure and excitation of even the simplest molecules is vastly more complex than atoms. Given the complicated structure, the Schrödinger equation of the system will be correspondingly complex, involving positions and moments of all constituents, both the nuclei and the electrons. Because the motion of the nuclei is so slow, the electrons make many cycles while the nuclei move to their new positions. This separation of the nuclear and electronic motion in molecular quantum mechanics is called the Born -Oppenheimer approximation.

Transitions in a molecule can therefore be put into three different categories according to different energies, W:

(a) electronic transitions with typical energies of a few eV – that are lines in the visual or UV regions of the spectrum;
(b) vibrational transitions caused by oscillations of the relative positions of nuclei with respect to their equilibrium positions. Typical energies are $0.1 - 0.01$ eV, corresponding to lines in the infrared region of the spectrum;
(c) rotational transitions caused by the rotation of the nuclei with typical energies of $\cong 10^{-3}$ eV corresponding to lines in the cm and mm wavelength range.

$$W^{\rm tot} = W^{\rm el} + W^{\rm vib} + W^{\rm rot} \,, \tag{150}$$

where W^{vib} and W^{rot} are the vibrational and rotational energies of the nuclei of the molecule, and W^{el} is the energy of the electrons. Under this assumption, the Hamiltonian is a sum of $W^{el} + W^{vib} + W^{rot}$. From quantum mechanics, the resulting wavefunction will be a product of the electronic, vibrational and rotational wavefunctions.

If we confine ourselves to the mm/sub-mm wavelength ranges, only transitions between different rotational levels and perhaps different vibrational levels (e.g. rotational transitions of SiO or HC_3N from vibrationally excited states) will be involved. This restriction results in a much simpler description of the molecular energy levels. Occasionally differences between geometrical arrangements of the nuclei result in a doubling of the energy levels. An example of such a case is the inversion doubling found for the Ammonia molecule.

12.2 Rotational Spectra of Diatomic Molecules

Because the effective radius of even a simple molecule is about 10^5 times the radius of the nucleus of an atom, the moment of inertia Θ_e of such a molecule is at least 10^{10} times that of an atom of the same mass. The kinetic energy of rotation is

$$H_{rot} = \frac{1}{2}\,\Theta_e\,\omega^2 = \boldsymbol{J}^2/2\,\Theta_e\,, \tag{151}$$

where $\boldsymbol{J}$ is the angular momentum. $\boldsymbol{J}$ is a quantity that cannot be neglected compared with the other internal energy states of the molecule, especially if the observations are made in the centimeter/millimeter/sub-mm wavelength ranges. (Note that $\boldsymbol{J}$ is the rotational quantum number, not the total orbital quantum number as in atomic physics.)

For a rigid molecule consisting of two nuclei A and B, the moment of inertia is

$$\Theta_e = m_A\,r_A^2 + m_B\,r_B^2 = m\,r_e^2\,, \tag{152}$$

where

$$\boldsymbol{r}_e = \boldsymbol{r}_A - \boldsymbol{r}_B \tag{153}$$

and

$$m = \frac{m_A\,m_B}{m_A + m_B}\,, \tag{154}$$

and

$$\boldsymbol{J} = \Theta_e\,\boldsymbol{\omega} \tag{155}$$

is the angular momentum perpendicular to the line connecting the two nuclei. For molecules consisting of three or more nuclei, similar, more complicated expressions can be obtained. Θ_e will depend on the relative orientation of the nuclei and will in general be a (three-axial) ellipsoid. In Eq. (155) values of Θ_e appropriate for the direction of $\boldsymbol{\omega}$ will then have to be used.

This solution of the Schrödinger equation then results in the *eigenvalues* for the rotational energy

$$E_{\rm rot} = W(J) = \frac{\hbar^2}{2\,\Theta_e}\,J(J+1)\,, \tag{156}$$

where J is the quantum number of angular momentum, which has integer values

$$J = 0, 1, 2, \ldots\,.$$

Equation (156) is correct only for a molecule that is completely rigid; for a slightly elastic molecule, r_e will increase with the rotational energy due to centrifugal stretching. (There is also the additional complication that even in the ground vibrational state there is still a zero point vibration; this will be discussed after the concept of centrifugal stretching is presented.) For centrifugal stretching, the rotational energy is modified to first order as:

$$E_{\rm rot} = W(J) = \frac{\hbar^2}{2\,\Theta_e}\,J(J+1) - hD\,[J(J+1)]^2\,. \tag{157}$$

Introducing the rotational constant

$$B_e = \frac{\hbar}{4\pi\,\Theta_e} \tag{158}$$

and the constant for centrifugal stretching, D, the pure rotation spectrum for electric dipole transitions $\Delta J = +1$ (emission) or $\Delta J = -1$ (absorption) is given by the following expression:

$$\nu(J) = \frac{1}{h}\,[W(J+1) - W(J)] = 2\,B_e\,(J+1) - 4D\,(J+1)^3\,. \tag{159}$$

Since D is positive, the observed line frequencies will be lower than those predicted on the basis of a perfectly rigid rotator. Typically, the size of D is about 10^{-5} of the magnitude of B_e for most molecules. In Fig. 15 we show a parameterized plot of the behavior of energy above ground and line frequency of a rigid rotor with and without the centrifugal distortion term. The function plotted vertically on the left, $E_{\rm rot}/B_e$, is proportional to the energy above the molecular ground state. This function is given by

$$E_{\rm rot}/B_e = 2\pi\hbar\,J(J+1) - 2\pi\hbar\,D/B_e\,[J(J+1)]^2\,, \tag{160}$$

while the levels on the right of Fig. 15 are given by the first term of Eq. (160) only. Directly below the energy level plots is a plot of the line frequencies for a number of transitions with quantum number J. The deviation between rigid rotor and actual frequencies becomes rapidly larger with increasing J, and in the sense that the actual frequencies are always lower than the frequencies predicted on the basis of a rigid rotor model. In Fig. 16 we show plots of the energies above ground state for a number of diatomic and triatomic linear molecules.

Allowed dipole radiative transitions will occur between different rotational states only if the molecule possesses a permanent electric dipole moment; that is, the molecule must be polar. Homonuclear diatomic molecules like H_2, N_2 or O_2 do not possess permanent electric dipole moments. Thus, they cannot undergo allowed transitions. This is one reason why it was so difficult to detect these species.

For molecules with permanent dipole moments, a classical picture of molecular line radiation can be used to determine the angular distribution of the radiation. In the plane of rotation, the dipole moment can be viewed as an antenna, oscillating as the molecule rotates. Classically, the acceleration of positive and negative charges gives rise to radiation whose frequency is that of the rotation frequency. For a dipole transition the most intense radiation occurs in the plane of rotation of the molecule. In the quantum mechanical model, the angular momentum is quantized, so that the radiation is emitted at discrete frequencies. Dipole radiative transitions occur with a change in the angular momentum quantum number of one unit, that is, $\Delta J = \pm 1$. The parity of the initial and final states must be opposite for dipole radiation to occur.

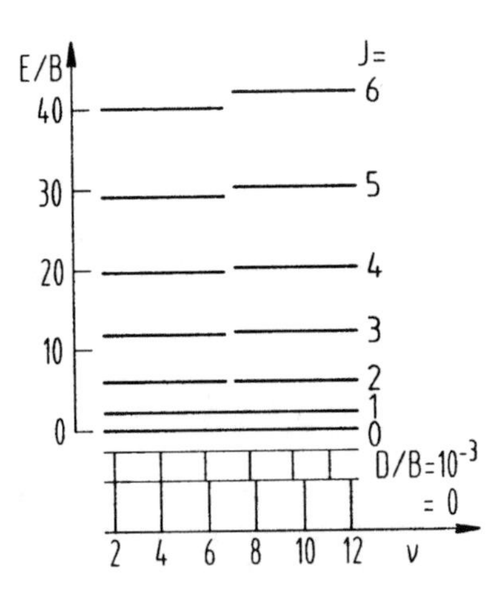

Fig. 15 A schematic plot of rotational energy levels for a rotor. The horizontal bars in the upper part represent the rotational energy levels for a rigid rotator (*right part*) and one deformed by centrifugal stretching with $D/B_e = 10^{-3}$ (*left part*). In fact, most molecules have $D/B_e \approx 10^{-5}$. The resulting line frequencies, ν, are shown in the lower part. The numbers next to ν refer to the J values. Figure taken from Rohlfs and Wilson (2004) [36]

12.3 Hyperfine Structure in Linear Molecules

The magnetic dipole or electric quadrupole of nuclei interact with electrons or other nuclei. These give rise to hyperfine structure in molecules such as HCN, HNC, and HC_3N. For example, the ^{14}N and Deuterium nuclei have spin $I = 1$ and thus a nonzero quadrupole moment. The hyperfine splitting of energy levels depends on the position of the nucleus in the molecule; the effect is smaller for HNC than for HCN. In general, the effect is of order of a few MHz, and decreases with increasing J. For nuclei with magnetic dipole moments, such as ^{13}C or ^{17}O, the hyperfine splitting is smaller. In the case of hyperfine structure, the total quantum number $\boldsymbol{F} = \boldsymbol{J} + \boldsymbol{I}$ is conserved. Allowed transitions obey the selection rule $\Delta F = \pm 1, 0$ but not $\Delta F = 0 \overrightarrow{\leftarrow} 0$.

12.4 Vibrational Transitions

If any of the nuclei of a molecule suffers a displacement from its equilibrium distance r_e, it will on release perform an oscillation about r_e. The Schrödinger equation for this is

$$\left(\frac{p^2}{2m} + P(r) \right) \psi^{\text{vib}}(x) = W^{\text{vib}} \psi^{\text{vib}}(x) , \tag{161}$$

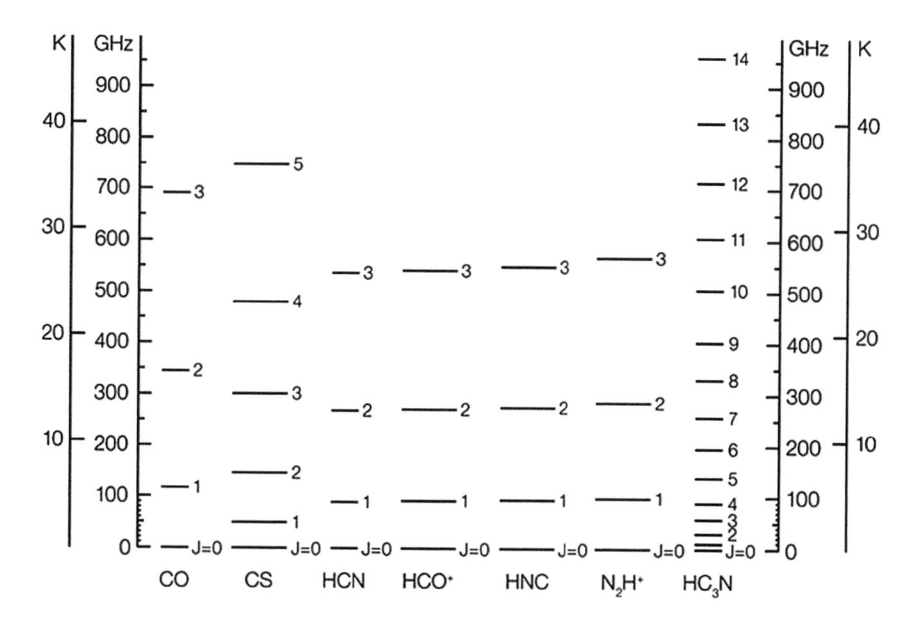

Fig. 16 Rotational energy levels of the vibrational ground states of some linear molecules which are commonly found in the interstellar medium. Figure taken from Wilson et al. (2013) [52]

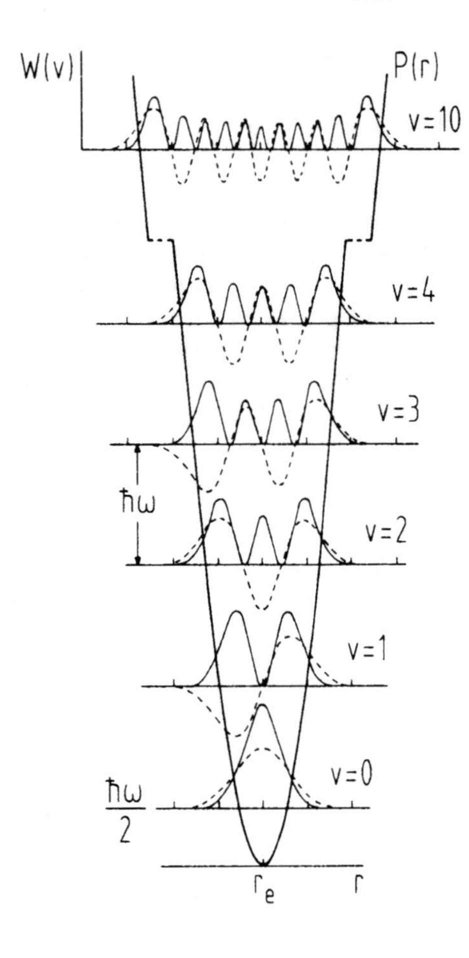

Fig. 17 Vibrational energy levels, eigenstates (dashed line) and probability densities (solid line) for a harmonic oscillator. Figure taken from Rohlfs and Wilson (2004) [36]

where $x = r - r_e$ and $P(r)$ is the potential function. If we have the simple harmonic approximation (Fig. 17) with the classical oscillation frequency

$$\omega = 2\pi\nu = \sqrt{\frac{k}{m}} = a\sqrt{\frac{2D_e}{m}}\,, \tag{162}$$

and Eq. (161) has the eigenvalue

$$W^{\mathrm{vib}} = W(v) = \hbar\omega\,(v + \tfrac{1}{2}) \tag{163}$$

with

$$v = 0, 1, 2, \ldots\,. \tag{164}$$

The solutions $\psi^{\rm vib}(x)$ can be expressed with the help of Hermite polynomials. For the same rotational quantum numbers, lines arising from transitions in different vibrational states, in a harmonic potential, are separated by a constant frequency interval.

For large x, the accuracy of the harmonic motion approximation is no longer sufficient, an empirical expression will have to be introduced into Eq. (161). The resulting differential equation can no longer be solved analytically, so numerical methods have to be used.

A molecule consisting of only two nuclei can vibrate only in one direction; it has only one vibrational mode. The situation is more complex for molecules with three or more nuclei. In this case, a multitude of various vibrational modes may exist, each of which will result in its own ladder of vibrational states, some of which may be degenerated. For a certain molecular vibrational state, there are many internal rotational states. Vibrational motions along the molecular axis can be and usually are hindered in the sense that these are subject to centrifugal forces, and thus must overcome an additional barrier.

It is possible to have transitions between rotational energy levels in a given vibrationally excited state. An example is the $J = 1 - 0$ rotational transition of the SiO molecule from the $v = 0$, 1 and 2 levels. The dipole moment in a vibrational state is usually the same as in the ground state. The dipole moment for a purely vibrational transition in the case of diatomic molecules is usually about 0.1 Debye. A more complex example is the polyatomic linear molecule HC_3N, for which a number of transitions have been measured in the ISM.

12.5 *Line Intensities of Linear Molecules*

In this section, we will give the details needed to relate the observed line intensities to column densities of the species emitting the transition. In the Born-Oppenheimer approximation, the total energy can be written as a sum of the electronic, vibrational and rotational energies in Eq. (150). In the line spectrum of a molecule, transitions between electronic, vibrational and rotational states are possible. We will restrict the discussion to rotational transitions, and in a few cases to vibrational transitions.

Computations of molecular line intensities proceed following the principles outlined in conjunction with the Einstein coefficients. The radial part of molecular wavefunctions is extremely complex. For any molecular or atomic system, the spontaneous transition probability, in s^{-1}, for a transition between an upper, u, and lower, l, level is given by the Einstein A coefficient A_{ul}. In the CGS system of units, A_{ul} is given by Eq. (134). Inserting numerical values, we have:

$$A_{ul} = 1.165 \times 10^{-11}\, \nu^3 |\mu_{ul}|^2 \,. \tag{165}$$

The units of the line frequency ν are GHz, and the units of μ are Debyes (i.e., 1 Debye $= 10^{-18}$ e.s.u.). Equation (165) is a completely general relation for *any* transition. The expression $|\mu|^2$ contains a term which depends on the integral over the angular part of the wavefunctions of the final and initial states; the radial part of the wavefunctions is contained in the value of the dipole moment, μ (this is usually determined from laboratory measurements). For dipole transitions between two rotational levels of a linear molecule, $J \rightleftarrows J+1$, there can be either absorption or emission. For the case of absorption, for a dipole moment $|\mu_{ul}|^2 = |\mu_J|^2$, we have:

$$|\mu_J|^2 = \mu^2 \, \frac{J+1}{2J+1} \qquad \text{for} \quad J \to J+1\,, \tag{166}$$

while for emission, the expression is given by:

$$|\mu_J|^2 = \mu^2 \, \frac{J+1}{2J+3} \qquad \text{for} \quad J+1 \to J\,, \tag{167}$$

where $\nu_{\rm r}$ is the spectral line frequency. Here, μ is the permanent electric dipole moment of the molecule. Table 2 contains parameters for a few species found in the interstellar medium.

After inserting Eq. (167) into Eq. (165), we obtain the expression for dipole emission between two levels of a linear molecule:

$$\boxed{A_J = 1.165 \times 10^{-11} \, \mu^2 \, \nu^3 \, \frac{J+1}{2J+3} \qquad \text{for} \quad J+1 \to J}\,, \tag{168}$$

where A is in units of s^{-1}, μ_J is in Debyes, and ν is in GHz. This expression is valid for a dipole transition in a linear molecule, from a level $J+1$ to J.

Inserting the expression for A in Eq. (129),the general relation between line optical depth, column density in a level l and excitation temperature, $T_{\rm ex}$, is:

$$\boxed{N_l = 93.5 \, \frac{g_l \, \nu^3}{g_u \, A_{ul}} \, \frac{1}{\left[1 - \exp(-4.80 \times 10^{-2} \nu / T_{\rm ex})\right]} \int \tau \, \mathrm{d}v}\,, \tag{169}$$

where the units for ν are GHz and the linewidths are in km s^{-1}. n is the local density in units of cm^{-3}, and $N = n \, l$ is the column density in cm^{-2}.

Although this expression appears simple, this is deceptive, since there is a dependence on $T_{\rm ex}$. The excitation process may cause $T_{\rm ex}$ to take on a wide range of values. If $T_{\rm ex}/\nu \gg 4.80 \times 10^{-2}$ K, the expression becomes:

$$\boxed{N_l = 1.94 \times 10^3 \, \frac{g_l \, \nu^2 \, T_{\rm ex}}{g_u \, A_{ul}} \int \tau \, \mathrm{d}v}\,. \tag{170}$$

Values for $T_{\rm ex}$ are difficult to obtain in the general case. Looking ahead a bit, for the $J=1 \to 0$ and $J=2 \to 1$ transitions, CO molecules are found to be almost

Table 2 Parameters of the commonly observed short cm/mm molecular lines

Chemical[a] formula	Molecule name	Transition	ν/GHz	E_u/K[b]	A_{ij}/s^{-1}[c]
H_2O	Ortho-water*	$J_{K_aK_c} = 6_{16} - 5_{23}$	22.235253	640	1.9×10^{-9}
NH_3	Para-ammonia	$(J, K) = (1, 1) - (1, 1)$	23.694506	23	1.7×10^{-7}
NH_3	Para-ammonia	$(J, K) = (2, 2) - (2, 2)$	23.722634	64	2.2×10^{-7}
NH_3	Ortho-ammonia	$(J, K) = (3, 3) - (3, 3)$	23.870130	122	2.5×10^{-7}
SiO	Silicon monoxide*	$J = 1 - 0, v = 2$	42.820587	3512	3.0×10^{-6}
SiO	Silicon monoxide*	$J = 1 - 0, v = 1$	43.122080	1770	3.0×10^{-6}
SiO	Silicon monoxide	$J = 1 - 0, v = 0$	43.423858	2.1	3.0×10^{-6}
CS	Carbon monosulfide	$J = 1 - 0$	48.990964	2.4	1.8×10^{-6}
DCO^+	Deuterated formylium	$J = 1 - 0$	72.039331	3.5	2.2×10^{-5}
SiO	Silicon monoxide*	$J = 2 - 1, v = 2$	85.640456	3516	2.0×10^{-5}
SiO	Silicon monoxide*	$J = 2 - 1, v = 1$	86.243442	1774	2.0×10^{-5}
$H^{13}CO^+$	Formylium	$J = 1 - 0$	86.754294	4.2	3.9×10^{-5}
SiO	Silicon monoxide	$J = 2 - 1, v = 0$	86.846998	6.2	2.0×10^{-5}
HCN	Hydrogen cyanide	$J = 1 - 0, F = 2 - 1$	88.631847	4.3	2.4×10^{-5}
HCO^+	Formylium	$J = 1 - 0$	89.188518	4.3	4.2×10^{-5}
HNC	Hydrogen isocyanide	$J = 1 - 0, F = 2 - 1$	90.663574	4.3	2.7×10^{-5}
N_2H^+	Diazenylium	$J = 1 - 0, F_1 = 2 - 1, F = 3 - 2$	93.173809	4.3	3.8×10^{-5}
CS	Carbon monosulfide	$J = 2 - 1$	97.980968	7.1	2.2×10^{-5}
$C^{18}O$	Carbon monoxide	$J = 1 - 0$	109.782182	5.3	6.5×10^{-8}
^{13}CO	Carbon monoxide	$J = 1 - 0$	110.201370	5.3	6.5×10^{-8}
CO	Carbon monoxide	$J = 1 - 0$	115.271203	5.5	7.4×10^{-8}
$H_2{}^{13}CO$	Ortho-formaldehyde	$J_{K_aK_c} = 2_{12} - 1_{11}$	137.449959	22	5.3×10^{-5}
H_2CO	Ortho-formaldehyde	$J_{K_aK_c} = 2_{12} - 1_{11}$	140.839518	22	5.3×10^{-5}
CS	Carbon monosulfide	$J = 3 - 2$	146.969049	14.2	6.1×10^{-5}
$C^{18}O$	Carbon monoxide	$J = 2 - 1$	219.560319	15.9	6.2×10^{-7}
^{13}CO	Carbon monoxide	$J = 2 - 1$	220.398714	15.9	6.2×10^{-7}
CO	Carbon monoxide	$J = 2 - 1$	230.538001	16.6	7.1×10^{-7}
CS	Carbon monosulfide	$J = 5 - 4$	244.935606	33.9	3.0×10^{-4}
HCN	Hydrogen cyanide	$J = 3 - 2$	265.886432	25.5	8.5×10^{-4}
HCO^+	Formylium	$J = 3 - 2$	267.557625	25.7	1.4×10^{-3}
HNC	Hydrogen isocyanide	$J = 3 - 2$	271.981067	26.1	9.2×10^{-4}

*Always found to be a maser transition

**Often found to be a maser transition

[a]If isotope not explicitly given, this is the most abundant variety, i.e., ^{12}C is C, ^{16}O is O, ^{14}N is N, ^{28}Si is Si, ^{32}S is S

[b]Energy of upper level above ground, in Kelvin

[c]Spontaneous transition rate, i.e., the Einstein A coefficient

always close to LTE, so it is possible to obtain estimates of T_K from Eq. (174). This result could be used in Eq. (170) if the transition is close to LTE. Expression (170) can be simplified even further if $\tau \ll 1$. Then, if the source fills the main beam (this is the usual assumption) the following relation holds:

$$T_{ex}\,\tau \cong T_{MB}\,, \tag{171}$$

where the term T_{MB} represents the main beam brightness temperature. In the general case, we will use T_b, which depends on source size. Inserting this in Eq. (170), we have:

$$\boxed{N_l = 1.94 \times 10^3 \frac{g_l\,\nu^2}{g_u\,A_{ul}} \int T_b\, dv}\,. \tag{172}$$

In this relation, T_{ex} appears nowhere. Thus, for an optically thin emission line, excitation plays *no role* in determining the column density in the energy levels giving rise to the transition. The units are as before; the column density, N_l, is an average over the telescope beam.

12.5.1 Total Column Densities of CO Under LTE Conditions

We apply the concepts developed in the last section to carbon monoxide, a simple molecule that is abundant in the ISM. Microwave radiation from this molecule is rather easily detectable because CO has a permanent dipole moment of $\mu =$ 0.112 Debye. CO is a diatomic molecule with a simple ladder of rotational levels spaced such that the lowest transitions are in the millimeter wavelength region. A first approximation of the abundance of the CO molecules can be obtained by a very standard LTE analysis of the CO line radiation; this is also fairly realistic since the excitation of low rotational transitions is usually close to LTE. Stable isotopes exist for both C and O and several isotopic species of CO have been measured in the interstellar medium; among these are $^{13}C^{16}O$, $^{12}C^{18}O$, $^{12}C^{17}O$, $^{13}C^{16}O$ and $^{13}C^{18}O$.

For the distribution of CO, we adopt the simplest geometry, that is, an isothermal slab which is much larger than the telescope beam. Then the solution (25) may be used. If we recall that a baseline is usually subtracted from the measured line profile, and that the 2.7 K microwave background radiation is present everywhere, the appropriate formula is

$$T_b(\nu) = T_0 \left(\frac{1}{e^{T_0/T_{ex}} - 1} - \frac{1}{e^{T_0/2.7} - 1} \right) (1 - e^{-\tau_\nu})\,, \tag{173}$$

where $T_0 = h\,\nu/k$. On the right side of Eq. (173) there are two unknown quantities: the excitation temperature of the line, T_{ex}, and the optical depth, τ_ν. If τ_ν is known, it is possible to solve for the column density N_{CO}, as in the case of the H I line at $\lambda = 21$ cm. But in the case of CO we meet the difficulty that lines of the most abundant isotope $^{12}\mathrm{C}\,^{16}\mathrm{O}$ always seem to be optically thick. It is therefore not possible to derive information about the CO column density from this line without a model for the molecular clouds. Here we give an analysis based on the measurement of weaker isotope lines of CO. This procedure can be applied if the following assumptions are valid:

- All molecules along the line of sight possess a uniform excitation temperature in the $J = 1 \to 0$ transition.
- The different isotopic species have the same excitation temperatures. Usually the excitation temperature is taken to be the kinetic temperature of the gas, T_{K}.
- The optical depth in the $^{12}\mathrm{C}\,^{16}\mathrm{O}$ $J = 1 \to 0$ line is large compared to unity.
- The optical depth in a rarer isotopomer transition, such as the $^{13}\mathrm{C}\,^{16}\mathrm{O}$ $J = 1 \to 0$ line is small compared to unity.
- The $^{13}\mathrm{CO}$ and CO lines are emitted from the same volume.

Given these assumptions, we have $T_{\mathrm{ex}} = T_{\mathrm{K}} = T$, where T_{K} is the kinetic temperature, which is the only parameter in the Maxwell-Boltzmann relation for the cloud in question. In the remainder of this section and in the following section we will use the expression T, since all temperatures are assumed to be equal. This is certainly *not* true in general. Usually, the molecular energy level populations are often characterized by *at least* the temperature, T_{ex}.

In general, the lines of $^{12}\mathrm{C}\,^{16}\mathrm{O}$ are optically thick. Then, in the absence of background continuum sources, the excitation temperature can be determined from the appropriate T_{b}^{12} of the optically thick $J = 1 - 0$ line of $^{12}\mathrm{C}\,^{16}\mathrm{O}$ at 115.271 GHz:

$$\boxed{T = 5.5 \Big/ \ln\left(1 + \frac{5.5}{T_{\mathrm{b}}^{12} + 0.82}\right)}\,. \qquad (174)$$

The optical depth of the $^{13}\mathrm{C}\,^{16}\mathrm{O}$ line at 110.201 GHz is obtained by solving Eq. (173) for

$$\boxed{\tau_0^{13} = -\ln\left[1 - \frac{T_{\mathrm{b}}^{13}}{5.3}\left\{\left[\exp\left(\frac{5.3}{T}\right) - 1\right]^{-1} - 0.16\right\}^{-1}\right]}\,. \qquad (175)$$

Usually the total column density is the quantity of interest. To obtain this for CO, one must sum over all energy levels of the molecule. This can be carried out for the LTE case in a simple way. For non-LTE conditions, the calculation is considerably more complicated. For more complex situations, statistical equilibrium or LVG models (Eq. (139)) are needed. In this section, we concentrate on the case of CO populations in LTE.

For CO, there is no statistical weight factor due to spin degeneracy. In a level J, the degeneracy is $2J + 1$. Then the fraction of the total population in a particular state, J, is given by:

$$N(J)/N(\text{total}) = \frac{(2J+1)}{Z} \exp\left[-\frac{h\,B_e J(J+1)}{kT}\right]. \tag{176}$$

Z is the sum over all states, or the Partition function. If vibrationally excited states are not populated, Z can be expressed as:

$$Z = \sum_{J=0}^{\infty} (2J+1) \exp\left[-\frac{h\,B_e\,J(J+1)}{k\,T}\right]. \tag{177}$$

The total population, $N(\text{total})$, is given by the measured column density for a specific level, $N(J)$, divided by the calculated fraction of the total population in this level:

$$N(\text{total}) = N(J)\,\frac{Z}{(2J+1)} \exp\left[\frac{h\,B_e\,J(J+1)}{k\,T}\right]. \tag{178}$$

This fraction is based on the assumption that all energy levels are populated under LTE conditions. For a temperature, T, the population will increase as $2J+1$, until the energy above the ground state becomes large compared to T. Then the negative exponential becomes the significant factor and the population will quickly decrease. If the temperature is large compared to the separation of energy levels, the sum can be approximated by an integral,

$$Z \approx \frac{k\,T}{h\,B_e} \quad \text{for} \quad h\,B_e \ll kT\,. \tag{179}$$

Here B_e is the rotation constant (Eq. (158)), and the molecular population is assumed to be characterized by a single temperature, T, so that the Boltzmann distribution can be applied. Applying Eq. (178) to the $J=0$ level, we can obtain the total column density of ^{13}CO from a measurement of the $J=1\rightarrow 0$ line of CO and ^{13}CO, using the partition function of CO, from Eqs. (179), and (169):

$$\boxed{N(\text{total})_{\text{CO}}^{13} = 3.0\times 10^{14}\,\frac{T\displaystyle\int \tau^{13}(v)\,\mathrm{d}v}{1-\exp\{-5.3/T\}}}\,. \tag{180}$$

It is often the case that in dense molecular clouds ^{13}CO is optically thick. Then we should make use of an even rarer substitution, C^{18}O. For the $J=1\rightarrow 0$ line of this substitution, the expression is exactly the same as Eq. (180). For the $J=2\rightarrow 1$ line, we obtain a similar expression, using Eq. (178):

$$\boxed{N(\text{total})^{13}_{\text{CO}} = 1.5 \times 10^{14} \frac{T\ \exp\{5.3/T\} \int \tau^{13}(v)\,\mathrm{d}v}{1 - \exp\{-10.6/T\}}} . \tag{181}$$

In both Eqs. (180) and (181), the beam averaged column density of carbon monoxide is in units of cm^{-2}, the line temperatures are in Kelvin, and the velocities, v, are in km s^{-1}. If the value of $T \gg 10.6$ or 5.3 K, the exponentials can be expanded to first order and then these relations become simpler.

In the limit of optically thin lines, integrals involving $\tau(v)$ are equal to the integrated line intensity $\int T_{\text{MB}}(v)\,\mathrm{d}v$, as mentioned before. However, there will be a dependence on T_{ex} in these relations because of the Partition function. The relation $T\,\tau(v) = T_{\text{MB}}(v)$ is only approximately true. However, optical depth effects can be eliminated to some extent by using the approximation

$$T \int_{-\infty}^{\infty} \tau(v)\,\mathrm{d}v \cong \frac{\tau_0}{1 - \mathrm{e}^{-\tau_0}} \int_{-\infty}^{\infty} T_{\text{MB}}(v)\,\mathrm{d}v . \tag{182}$$

This formula is accurate to 15% for $\tau_0 < 2$, and it always overestimates N when $\tau_0 > 1$. The formulas (174), (175), (180), and (181) permit an evaluation of the column density N^{13}_{CO} *only* under the assumption of LTE.

The most extensive and complete survey in the $J = 1 - 0$ line of ^{13}CO was carried out by the Boston University FCRAO group [22]. This covers the inner part of the northern galaxy with full sampling; there are nearly 2×10^6 spectra.

For other linear molecules, the expressions for the dipole moments and the partition functions are similar to that for CO and the treatment is similar. There is one very important difference however. The simplicity in the treatment of the CO molecule arises because of the assumption of LTE or near-LTE conditions. This may not be the case for molecules such as HCN or CS, since these species have dipole moments of the order of 2 to 3 Debye. Thus populations of high J levels (which have faster spontaneous decay rates) may have populations lower than predicted by LTE calculations. Such populations are said to be *subthermal*, because the excitation temperature characterizing the populations would be $T_{\text{ex}} < T_{\text{K}}$.

12.6 Symmetric Top Molecules

12.6.1 Energy Levels

Symmetric and asymmetric top molecules are vastly more complex than linear molecules. The rotation of a rigid molecule with an arbitrary shape can be considered to be the superposition of three free rotations about the three principal axes of the inertial ellipsoid. Depending on the symmetry of the molecule these principal axes can all be different: in that case the molecule is an asymmetric top. If two principal axes are equal, the molecule is a symmetric top. If all three principal axes are equal, it is a spherical top. In order to compute the angular parts of the wavefunction, the proper Hamiltonian operator must be solved in the Schrödinger equation and the stationary state eigenvalues determined.

In general, for any rigid rotor asymmetric top molecule in a stable state, the total momentum $\boldsymbol{J}$ will remain constant with respect to both its absolute value and its direction. As is known from atomic physics, this means that both $(\boldsymbol{J})^2$ and the projection of $\boldsymbol{J}$ into an arbitrary but fixed direction, for example J_z, remain constant. If the molecule is in addition symmetric, the projection of $\boldsymbol{J}$ on the axis of symmetry will be constant also.

For the *symmetric top molecule*, $\boldsymbol{J}$ is inclined with respect to the axis of symmetry z. Then the figure axis z will precess around the direction $\boldsymbol{J}$ forming a constant angle with it, and the molecule will simultaneously rotate around the z axis with the constant angular momentum J_z. From the definition of a symmetric top, $\Theta_x = \Theta_y$. Taking $\Theta_x = \Theta_y = \Theta_\perp$ and $\Theta_z = \Theta_\parallel$, we obtain a Hamiltonian operator:

$$H = \frac{J_x^2 + J_y^2}{2\,\Theta_\perp} + \frac{J_z^2}{2\,\Theta_\parallel} = \frac{\boldsymbol{J}^2}{2\,\Theta_\perp} + J_z^2 \cdot \left(\frac{1}{2\,\Theta_\parallel} - \frac{1}{2\,\Theta_\perp} \right) . \tag{183}$$

Its eigenvalues are:

$$W(J,K) = J(J+1)\,\frac{\hbar^2}{2\,\Theta_\perp} + K^2\,\hbar^2 \left(\frac{1}{2\,\Theta_\parallel} - \frac{1}{2\,\Theta_\perp} \right) , \tag{184}$$

where K^2 is the eigenvalue from the operator J_z^2 and $J^2 = J_x^2 + J_y^2 + J_z^2$ is the eigenvalue from the operator $J_x^2 + J_y^2 + J_z^2$.

The analysis of linear molecules is a subset of that for symmetric molecules. For linear molecules, $\Theta_\parallel \to 0$ so that $1/(2\,\Theta_\parallel) \to \infty$. Then finite energies in Eq. (184) are possible only if $K = 0$. For these cases the energies are given by Eq. (156). For symmetric top molecules each eigenvalue has a multiplicity of $2J + 1$ with:

$$J = 0, 1, 2, \ldots \quad K = 0, \pm 1, \pm 2, \cdots \pm J\ . \tag{185}$$

From Eq. (184), the energy is independent of the sign of K, so levels with the same J and absolute value of K coincide. Then levels with $K > 0$ are doubly degenerate.

It is usual to express $\dfrac{\hbar}{4\pi\,\Theta_\perp}$ as B, and $\dfrac{\hbar}{4\pi\,\Theta_\parallel}$ as C. The units of these rotational constants, B and C are usually either MHz or GHz. Then Eq. (184) becomes

$$W(J,K)/h = B\,J(J+1) + K^2\,(C - B)\ . \tag{186}$$

12.6.2 Spin Statistics

In the case of molecules containing identical nuclei, the exchange of such nuclei, for example by the rotation about an axis, has a spectacular effect on the selection rules. Usually there are no interactions between electron spin and rotational motion. Then the total wavefunction is the product of the spin and rotational wavefunctions. Under an interchange of fermions, the total wave function must be antisymmetric (these identical nuclei could be protons or have an uneven number of nucleons). The symmetry of the spin wavefunction of the molecule will depend on the relative orientation of the spins. If the spin wavefunction is symmetric, this is the *ortho*-modification of the molecule; if antisymmetric it is the *para* -modification. In thermal equilibrium in the ISM, collisions with the exchange of identical particles will change one modification into the other only very slowly, on time scales of $> 10^6$ years. This could occur much more quickly on grain surfaces, or with charged particles. If the exchange is slow, the ortho and para modifications of a particular species behave like different molecules; a comparison of ortho and para populations might give an estimate of temperatures in the distant past, perhaps at the time of molecular formation.

For the H_2 molecule, the symmetry of the rotational wavefunction depends on the total angular momentum J as $(-1)^J$. In the $J = 0$ state the rotational wavefunction is symmetric. However, the total wavefunction must be antisymmetric since protons are fermions. Thus, the $J = 0, 2, 4$, etc., rotational levels are para-H_2, while the $J = 1, 3, 5$, etc., are ortho-H_2. Spectral lines can connect only one modification. In the case of H_2, dipole rotational transitions are not allowed, but quadrupole rotational transitions ($\Delta J = \pm 2$) are. Thus, the 28 μm line of H_2 connects the $J = 2$ and $J = 0$ levels of para -H_2. Transitions between the ground and vibrational states are also possible.

Finally, as a more complex example of the relation of identical nuclei, we consider the case of three identical nuclei. This is the case for NH_3, CH_3CN and CH_3C_2H. Exchanging two of the nuclei is equivalent to a rotation by 120°. An exchange as was used for the case of two nuclei would not, in general, lead to a suitable symmetry. Instead combinations of spin states must be used. These lead to the result that the ortho to para ratio is two to one if the identical nuclei are protons. That is, NH_3, CH_3CN or CH_3C_2H the ortho form has $S(J, K) = 2$, while the para form has $S(J, K) = 1$. In summary, the division of molecules with identical nuclei into ortho and para species determines selection rules for radiative transitions and also rules for collisions (see e.g. [48]).

12.6.3 Hyperfine Structure

For symmetric top molecules, the simplest hyperfine spectra is found for the inversion doublet transitions of NH_3. Since both the upper and lower levels have the same quantum numbers (J, K), there will be 5 groups of hyperfine components separated by a few MHz. Because of interactions between the spins of H nuclei, there will be

Table 3 Intensities of satellite groups relative to the Main Component (see [36])

(J, K)	(1, 1)	(2, 2)	(3, 3)	(4, 4)	(5, 5)	(6, 6)	(2, 1)
I_{inner}	0.295	0.0651	0.0300	0.0174	0.0117	0.0081	0.0651
I_{outer}	0.238	0.0628	0.0296	0.0173	0.0114	0.0081	0.0628

an additional splitting, within each group, of the order of a few kHz. In Table 3 we give the relative intensities of the NH_3 satellites for the case of low optical depth and LTE. For a molecule such as OH, one of the electrons is unpaired. The interaction of the nuclear magnetic moment with the magnetic moment of an unpaired electron is described as magnetic hyperfine structure. This splits a specific line into a number of components. In the case of the OH molecule, this interaction gives rise to a hyperfine splitting of the energy levels, in addition to the much larger Λ doubling. Together with the Λ doublet splitting, this gives rise to a quartet of energy levels in the OH ground state. Transitions between these energy levels produces the four ground state lines of OH at 18 cm wavelength (see Fig. 18).

NH_3 is an example of an oblate symmetric top molecule commonly found in the ISM. A diagram of the lower energy levels of NH_3 are shown in Fig. 19. A prolate top molecule has a cigar-like shape. Then A replaces C, and $A > B$. The energy-level diagrams for prolate symmetric top molecules found in the ISM, such as CH_3CCH and CH_3CN, follow this rule. However, since these molecules are much heavier than NH_3, the rotational transitions give rise to lines in the millimeter wavelength range.

Differences in the orientation of the nuclei can be of importance. If a reflection of all particles about the center of mass leads to a configuration which cannot be obtained by a rotation of the molecule, so these reflections represent two different states. For NH_3, we show this situation in the upper part of Fig. 19. Then there are two separate, degenerate states which exist for each value of (J, K) for $J \geq 1$ (the $K = 0$ ladder is an exception because of the Pauli principle). These states are doubly degenerated as long as the potential barrier separating the two configurations is infinitely high. However, in molecules such as NH_3 the two configurations are separated only by a small potential barrier. This gives rise to a measurable splitting of the degenerate energy levels, which is referred to as inversion doubling. For NH_3, transitions between these inversion doublet levels are caused by the quantum mechanical tunneling of the nitrogen nucleus through the plane of the three protons. The wavefunctions of the two inversion doublet states have opposite parities, so that dipole transitions are possible. Thus dipole transitions *can* occur between states with the same (J, K) quantum numbers. The splitting of the (J, K) levels for NH_3 shown in Fig. 19 is exaggerated; the inversion transitions give rise to spectral lines in the wavelength range near 1 cm. For CH_3CCH or CH_3CN, the splitting caused by inversion doubling is very small, since the barrier is much higher than for NH_3.

The direction of the dipole moment of symmetric top molecules is parallel to the K axis. Spectral line radiation can be emitted only by a changing dipole moment. Since radiation will be emitted perpendicular to the direction of the dipole moment, there

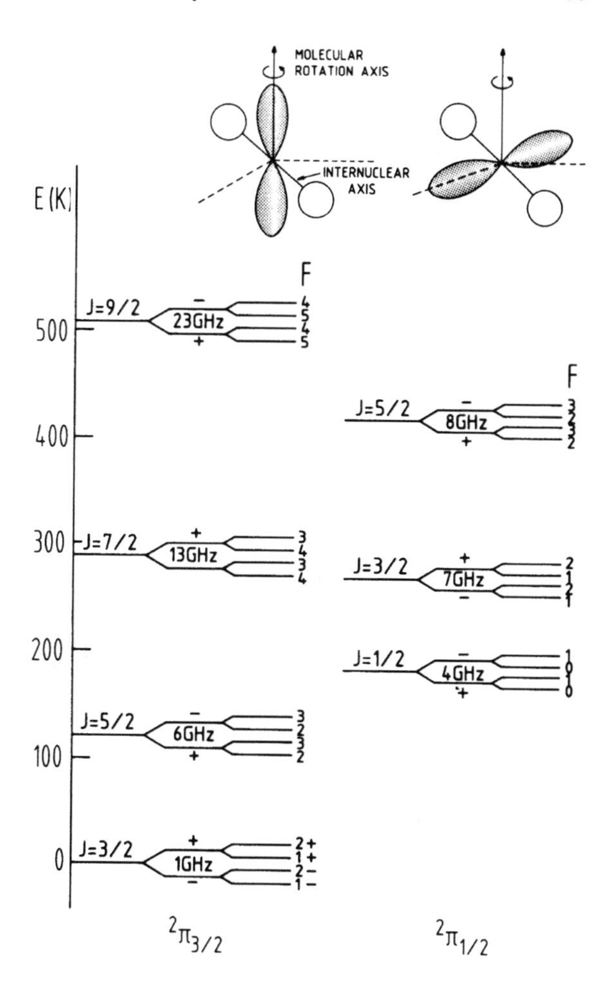

Fig. 18 The lower energy levels of OH showing Λ doubling. F is the total angular momentum, including electron spin, while J is the rotational angular momentum due to the nuclear motion. The quantum number F includes hyperfine splitting of the energy levels. The parities of the states are also shown under the symbol F. The Λ doubling causes a splitting of the J states. In the sketch of the OH molecule, the shaded regions represent the electron orbits in the Λ state. The two unshaded spheres represent the O and H nuclei. The configuration shown on the left has the higher energy. Figure taken from Rohlfs and Wilson (2004) [36]

can be no radiation along the symmetry axis. Thus the K quantum number *cannot* change in dipole radiation, so allowed dipole transitions cannot connect different K ladders. The different K ladders are connected by octopole radiative transitions which require $\Delta K = \pm 3$. These are very slow, however, and collisions are far more likely to cause an exchange of population between different K ladders. This is used to estimate T_K from the ratio of populations of different (J, K) states in symmetric top molecules.

12.6.4 Line Intensities and Column Densities

The extension of this analysis to symmetric top molecules is only slightly more complex. The dipole moment for an allowed transition between energy levels $J + 1, K$ and J, K for a symmetric top such as CH_3CN or CH_3C_2H is

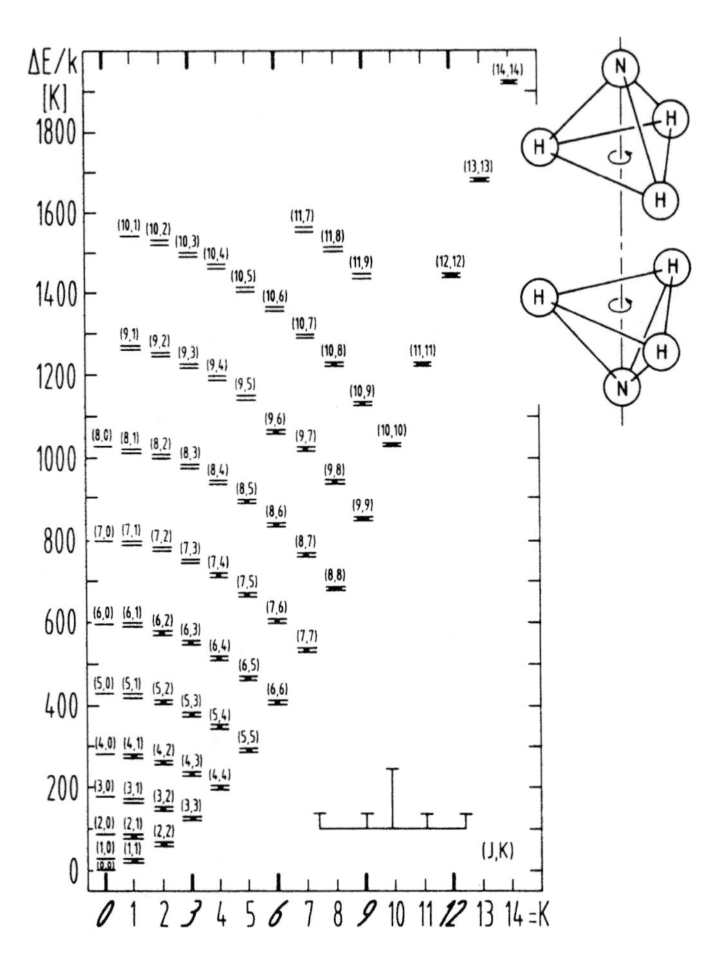

Fig. 19 The energy-level diagram of the vibrational ground state of NH_3, an oblate symmetric top molecule. Ortho-NH_3 has $K = 0, 3, 6, 9, \cdots$, while para-NH_3 has all other K values (see text). Rotational transitions with $\Delta J = 1$, $\Delta K = 0$, give rise to lines in the far IR. This molecule also has transitions with $\Delta J = 0$, $\Delta K = 0$ between inversion doublet levels. The interaction of the nuclear spin of ^{14}N with the electrons causes quadrupole hyperfine structure. In the $\Delta J = 0$, $\Delta K = 0$ transitions, the line is split into 5 groups of components. A sketch of the structure of the groups of hyperfine components of the $(J, K) = (1, 1)$ inversion doublet line is indicated in the lower right; the separation is of order of MHz. In the upper right is a sketch of the molecule before and after an inversion transition, which gives rise to a 1.3 cm photons. Figure taken from Rohlfs and Wilson (2004) [36]

$$|\mu_{JK}|^2 = \mu^2 \, \frac{(J+1)^2 - K^2}{(J+1)(2J+3)} \qquad \text{for} \quad (J+1, K) \to (J, K)\,. \tag{187}$$

For these transitions, $J \geq K$ always.

For NH_3, the most commonly observed spectral lines are the inversion transitions at 1.3 cm between levels (J, K) and (J, K). The dipole moment is

$$|\mu_{JK}|^2 = \mu^2 \frac{K^2}{J(J+1)} \quad \text{for} \quad \Delta J = 0,\ \Delta K = 0\,. \tag{188}$$

When these relations are inserted in Eq. (165), the population of a specific level can be calculated following Eq. (169). If we follow the analysis used for CO, we can use the LTE assumption to obtain the entire population

$$N(\text{total}) = N(J, K) \frac{Z}{(2J+1)\,S(J, K)} \exp\left[\frac{W(J, K)}{k\,T}\right], \tag{189}$$

where W is the energy of the level above the ground state, and the nuclear spin statistics are accounted for through the factor $S(J, K)$ for the energy level corresponding to the transition measured. For symmetric top molecules, we have, using the expression for the energy of the level in Eq. (186),

$$N(\text{total}) = \frac{Z\,N(J, K)}{(2J+1)\,S(J, K)} \exp\left[\frac{B\,J(J+1) + (C-B)K^2}{k\,T}\right]. \tag{190}$$

For prolate tops, A replaces C in Eq. (190), and in Eqs. (191)–(195). If we sum over all energy levels, we obtain $N(\text{total})$ and the partition function, Z, in the following:

$$Z = \sum_{J=0}^{\infty} \sum_{K=0}^{K=J} (2J+1)\,S(J, K) \exp\left[-\frac{B\,J(J+1) + (C-B)K^2}{k\,T}\right]. \tag{191}$$

If the temperature is large compared to the spacing between energy levels, one can replace the sums by integrals, so that:

$$Z \approx \sqrt{\frac{\pi (k\,T)^3}{h^3\,B^2\,C}}\,. \tag{192}$$

If we assume that $h\nu \ll kT$, use CGS units for the physical constants, and GHz for the rotational constants A, B and C, the partition function, Z, becomes

$$Z \approx 168.7 \sqrt{\frac{T^3}{B^2\,C}}\,. \tag{193}$$

Substituting into Eq. (190), we have:

$$N(\text{total}) = N(J, K) \frac{168.7 \sqrt{\frac{T^3}{B^2\,C}}}{(2J+1)\,S(J, K)} \exp\left[\frac{W(J, K)}{k\,T}\right]. \tag{194}$$

Here, $N(J, K)$ can be calculated from Eq. (169) or Eq. (170), using the appropriate expressions for the dipole moment (Eq. (188)) in the Einstein A coefficient relation

Eq. (165). W is the energy of the level above the ground state. In the ISM, ammonia inversion lines up to $(J, K) = (18, 18)$ have been detected [51].

We now consider a situation in which the NH_3 population is *not* thermalized. This is typically the case for dark dust clouds. We must use some concepts presented in the next few sections for this analysis. If $n(H_2) \sim 10^4$ cm^{-3}, and the infrared field intensity is small, a symmetric top molecule such as NH_3 can have a number of excitation temperatures. The excitation temperatures of the populations in doublet levels are usually between 2.7 K and T_K. The rotational temperature, T_{rot}, which describes populations for metastable levels ($J = K$) in different K ladders, is usually close to T_K. This is because radiative transitions between states with a different K value are forbidden to first order. The excitation temperature which describes the populations with different J values within a given K ladder will be close to 2.7 K, since radiative decay with $\Delta K = 0$, $\Delta J = 1$ is allowed. Then the non-metastable energy levels ($J > K$) are not populated. In this case, Z is simply given by the sum over the populations of metastable levels:

$$Z(J = K) = \sum_{J=0}^{\infty} (2J + 1)\, S(J, K = J)\, \exp\left[-\frac{BJ(J + 1) + (C - B)J^2}{kT}\right] . \quad (195)$$

For the NH_3 molecule in dark dust clouds, where $T_K = 10$ K and $n(H_2) = 10^4$ cm^{-3}, we can safely restrict the sum to the three lowest metastable levels:

$$Z(J = K) \approx N(0, 0) + N(1, 1) + N(2, 2) + N(3, 3) . \quad (196)$$

Substituting the values for NH_3 metastable levels:

$$Z(J = K) \approx N(1, 1)\left[\frac{1}{3}\exp\left(\frac{23.1}{kT}\right) + 1 + \frac{5}{3}\exp\left(-\frac{41.2}{kT}\right) + \frac{14}{3}\exp\left(-\frac{99.4}{kT}\right)\right] . \quad (197)$$

For NH_3 we have given two extreme situations: in the first case, described by Eq. (197), is a low-density cloud for which only the few lowest metastable levels are populated. The second case is the LTE relation, given in Eq. (194). This represents a cloud in which the populations of the molecule in question are thermalized. More complex are those situations for which the populations of some of the levels are thermalized, and others not. Such a situation requires the use of a statistical equilibrium or an LVG model.

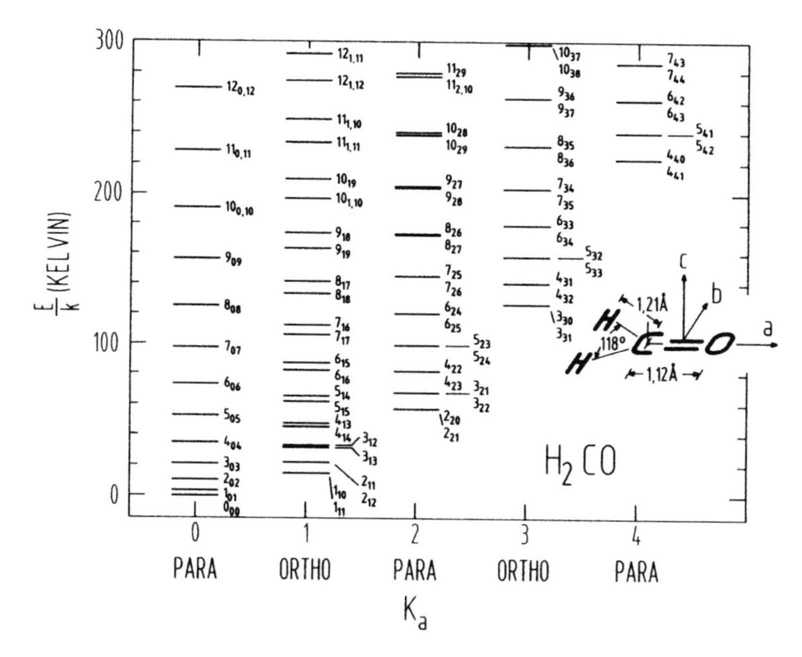

Fig. 20 Energy-level diagram of formaldehyde, H_2CO. This is a planar asymmetric top, but the asymmetry is very small. The energy-level structure is typical of an almost prolate symmetric top molecule. In the lower right is a sketch of the structure of the molecule. Figure taken from Mangum and Wootten (1993) [28]

12.7 Asymmetric Top Molecules

12.7.1 Energy Levels

For an *asymmetric top molecule*, there are no internal molecular axes with a time-invariable component of angular momentum. So only the total angular momentum is conserved and we have only J as a good quantum number. The moments of inertia about each axis are different; the rotational constants are referred to as A, B and C, with $A > B > C$. The prolate symmetric top ($B = C$) or oblate symmetric top ($B = A$) molecules can be considered as the limiting cases. But neither the eigenstates nor the eigenvalues are easily expressed in explicit form. Each of the levels must be characterized by three quantum numbers. One choice is $J_{K_a K_c}$, where J is the total angular momentum, K_a is the component of J along the A axis and K_c is the component along the C axis. If the molecule is a prolate symmetric top, J and K_a are good quantum numbers; if the molecule were an oblate symmetric top, J and K_c would be good quantum numbers. Intermediate states are characterized by a superposition of the prolate and oblate descriptions. In Fig. 20, we show the energy-level diagram for the lower levels of H_2CO, which is almost a prolate symmetric top molecule with the dipole moment along the A axis. Since radiation must be

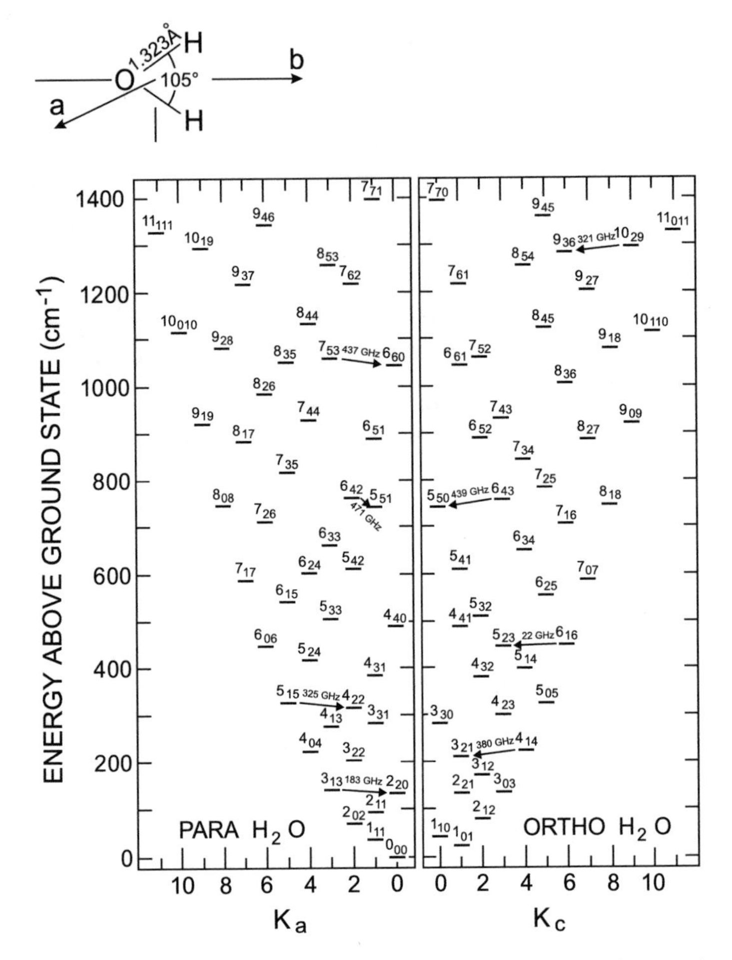

Fig. 21 Energy-level diagrams for ortho- and para-H_2O. This is an asymmetric top molecule, with the dipole moment along the *B* axis, that is the axis with an intermediate moment of inertia. Because of the two identical nuclei, the energy-level diagram is split into ortho and para, that show almost no interaction under interstellar conditions. The transitions marked by arrows are well-known masers [8, 36]

emitted perpendicular to the direction of the dipole moment, for H_2CO there can be no radiation emitted along the *A* axis, so the quantum number K_a will not change in radiative transitions.

12.7.2 Spin Statistics and Selection Rules

The case of a planar molecule with two equivalent nuclei, such as H_2CO, shows is a striking illustration of these effects (see Fig. 20). The dipole moment lies along the A axis. A rotation by 180° about this axis will change nothing in the molecule, but will exchange the two protons. Since the protons are fermions, this exchange must lead to an antisymmetric wavefunction. Then the symmetry of the spin wave function and the wave function describing the rotation about the A axis must be antisymmetric. The rotational symmetry is $(-1)^{K_a}$. If the proton spins are parallel, that is ortho-H_2CO, then the wave function for K_a must be anti-symmetric, or K_a must take on an odd value. If the proton spins are anti-parallel, for para-H_2CO, K_a must have an even value (Fig. 20). For ortho -H_2CO, the parallel spin case, there are three possible spin orientations. For para-H_2CO, there is only one possible orientation, so the ratio of ortho-to-para states is three. Such an effect is taken into account in partition functions (177) by spin degeneracy factors, which are denoted by the symbol $S(J, K)$. For ortho-H_2CO, $S(J, K) = 3$, for para-H_2CO, $S(J, K) = 1$. This concept will be applied in Sect. 12.7.3.

Allowed transitions can occur only between energy levels of either the ortho or the para species. For example, the 6 cm H_2CO line is emitted from the ortho modification only (see Fig. 20). Another example is the interstellar H_2O maser line at $\lambda = 1.35$ cm which arises from ortho-H_2O (see Fig. 21). The H_2O molecule is a more complex case, since the dipole moment is along the B axis. Then in a radiative transition, both K_a and K_c must change between the initial and final state.

12.7.3 Line Intensities and Column Densities

For asymmetric molecules the moments of inertia for the three axes are all different; there is no symmetry, so three quantum numbers are needed to define an energy level. The relation between energy above the ground state and quantum numbers is given in Appendix IV of [48] or in databases for specific molecules (see figure captions for references). The relation for the dipole moment of a specific transition is more complex; generalizing from Eq. (165), we have for a spontaneous transition from a higher state, denoted by u, to a lower state, denoted by l:

$$A = 1.165 \times 10^{-11}\, \nu_{\rm x}^3\, \mu_x^2\, \frac{\mathcal{S}(u; l)}{2J' + 1}\,. \tag{198}$$

This involves a dipole moment in a direction x. As before, the units of ν are GHz, and the units of μ are Debyes. The value of the quantum number J' refers to the lower state. The expression $\mathcal{S}(u; l)$ is the angular part of the dipole moment between the initial and final state. The dipole moment can have a direction which is *not* along a single axis. In this case there are different values of the dipole moment along different molecular axes. In contrast, for symmetric top or linear molecules, there is *a* dipole moment for rotational transitions. Methods to evaluate transition probabilities for asymmetric molecules are discussed at length in [48]. There is a table of $\mathcal{S}(u; l)$ in their Appendix V. From the expression for the Einstein A coefficient, the column density in a given energy level can be related to the line intensity by Eq. (169). Following the procedures used for symmetric top molecules, we can use a relation similar to Eq. (190) to sum over all levels, using the appropriate energy, W, of the level $J_{K_a K_c}$ above the ground state and the factor $S(J_{K_a K_c})$ for spin statistics:

$$N(\text{total}) = N(J_{K_a K_c}) \frac{Z}{(2J+1)\, S(J_{K_a K_c})} \exp\left(\frac{W}{k\,T}\right). \tag{199}$$

If the populations are in LTE, one can follow a process similar to that used to obtain Eq. (193). Then we obtain the appropriate expression for the partition function:

$$Z = 168.7 \sqrt{\frac{T^3}{A\,B\,C}}. \tag{200}$$

When combined with the Boltzmann expression for a molecule in a specific energy level, this gives a simple expression for the fraction of the population in a specific rotational state if LTE conditions apply:

$$N(\text{total}) \approx N(J_{K_a K_c}) \frac{168.7 \sqrt{\frac{T^3}{A\,B\,C}}}{(2J+1)\, S(J_{K_a K_c})} \exp\left(\frac{W}{k\,T}\right). \tag{201}$$

In this expression, $S(J_{K_a K_c})$ accounts for spin statistics for energy level $J_{K_a K_c}$, and A, B and C are the molecular rotational constants in GHz. W, the energy of the level above the ground state, and T, the temperature, are given in Kelvin. Given the total molecular column density and the value of T, the feasibility of detecting a specific line can be obtained when the appropriate A coefficient value is inserted into Eq. (169) or (170).

As pointed out in connection with NH_3, T needs not be T_K. In reality, a number of different values of T may be needed to describe the populations. We will investigate the influence of excitation conditions on molecular populations and observed line intensities next.

Two important interstellar molecules are H_2CO and H_2O. Here we summarize the dipole selection rules. Rotating the molecule about the axis along the direction of the dipole moment, we effectively exchange two identical particles. If these are fermions, under this exchange the total wavefunction must be antisymmetric. For H_2CO, in Sect. 12.6.2 we reviewed the spin statistics. Since the dipole moment is along the A axis, a dipole transition must involve a change in the quantum numbers along the B or C axes. From Fig. 20, the $K_a = 0$ ladder is para-H_2CO, so to have a total wavefunction which is antisymmetric, one must have a space wavefunction which is symmetric. For a dipole transition, the parities of the initial and final states must have different parities. This is possible if the C quantum number changes. For H_2O, the dipole moment is along the B axis (see Fig. 21). In a dipole transition, the quantum number for the B direction will not change. For ortho-H_2O, the spin wavefunction is symmetric, so the symmetry of the space wavefunction must be antisymmetric. In general, this symmetry is determined by the product of K_a and K_c. For ortho-H_2O, this must be $K_a K_c =$ (odd) (even), i.e., *oe*, or *eo*. For allowed transitions, one can

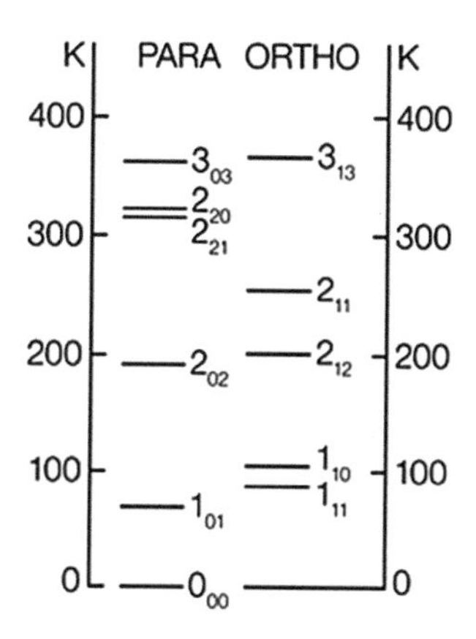

Fig. 22 The few lowest energy levels of the H_2D^+ molecule. This is a planar, triangular-shaped asymmetric top. Figure taken from Gerlich et al. (2006) [12]

have $oe - eo$ or $eo - oe$. For para-H_2O, the rule is $oo - ee$ or $ee - oo$. Clearly, H_2S follows the selection rules for H_2O. These rules will be different for SO_2, since the exchanged particles are bosons. More exotic are D_2CO, ND_3 and D_2O.

The species H_3^+ has the shape of a planar triangle. It is a key to ion-molecule chemistry, but has no rotational transitions because of its symmetry. The deuterium isotopomer, H_2D^+, is an asymmetric top molecule with a permanent dipole moment. The spectral line from the 1_{10}-1_{11} levels, an ortho species, was found at 372.421 GHz. A far infrared absorption line from the 2_{12}-1_{11} levels was also detected. We show an energy-level diagram in Fig. 22. The doubly deuterated species, D_2H^+,has been detected in the 1_{10}-1_{01} line at 691.660 GHz from the para species (see [49]).

12.8 Electronic Angular Momentum

In many respects the description of electronic angular momentum is similar to that of atomic fine structure as described by Russell-Saunders coupling. Each electronic state is designated by the symbol ${}^{2S+1}\Lambda_{\Omega}$, where $2S + 1$ is the multiplicity of the state with S the electron spin and Λ the projection of the electronic orbital angular momentum on the molecular axis in units of $\hbar$. The molecular state is described as Σ, Π, Δ etc., according to whether $\Lambda = 0, 1, 2, \ldots$.

Σ is the projection of the electron spin angular momentum on the molecular axis in units of $\hbar$ (not to be confused with the symbol Σ, for $\Lambda = 0$). Finally, Ω is the total electronic angular momentum. For the Hund coupling case A, $\Omega = |\Lambda + \Sigma|$ (see e.g. [36]).

Since the frequencies emitted or absorbed by a molecule in the optical range are due to electronic state changes, many of the complications found in optical spectra are not encountered when considering transitions in the cm and mm range. However, the electronic state does affect the vibrational and rotational levels even in the radio range. For most molecules, the ground state has zero electronic angular momentum, that is, a singlet sigma, ${}^1\Sigma$ state. For a small number of molecules such as OH,

CH, C_2H, or C_3H, this is not the case; these have ground state electronic angular momentum. Because of this fact, the rotational energy levels experience an additional energy-level splitting, which is Λ doubling. This is a result of the interaction of the rotation and the angular momentum of the electronic state. This splitting causes the degenerate energy levels to separate. This splitting can be quite important for Π states; for Δ and higher states it is usually negligible. The OH molecule is a prominent example for this effect. Semi-classically, the Λ doubling of OH can be viewed as the difference in rotational energy of the (assumed rigid) diatomic molecule, when the electronic wave function is oriented with orbitals in a lower or higher moment of inertia state. We show a sketch of this in the upper part of Fig. 18. Since the energy is directly proportional to the total angular momentum quantum number and inversely proportional to the moment of inertia, the molecule shown on the left has higher energy than the one shown on the right.

There are also a few molecules for which the orbital angular momentum is zero, but the electron spins are parallel, so that the total spin is unity. These molecules have triplet sigma $^3\Sigma$ ground states. The most important astrophysical example is the SO molecule; another species with a triplet $^3\Sigma$ ground state is O_2. These energy levels are characterized by the quantum number, N, and the orbital angular momentum quantum number J. The most probable transitions are those within a ladder with $\Delta J = \Delta N = \pm 1$, but there can be transitions across N ladders. As with the OH molecule, some states of SO are very sensitive to magnetic fields. One could then use the Zeeman effect to determine the magnetic field strength. This may be difficult, since a 1 μGauss field will cause a line splitting of only about 1 Hz in the linear polarization. Even so, measurements of the polarization of the $J_N = 1_0 - 0_1$ line of SO may allow additional determinations of interstellar magnetic fields [36].

12.8.1 Molecules with Hindered Motions

The most important hindered motion involve quantum mechanical tunneling; such motions cannot occur in classical mechanics because of energy considerations. A prime example of this phenomenon is the motion of the hydrogen atom attached to oxygen in methanol CH_3OH. This H atom can move between 3 positions between the three H atoms in the CH_3 group. Another example is the motion of the CH_3 group in methyl formate CH_3COOH. These are dependent on the energy. At low energy these motions do not occur, while at larger energies they are more important. For both methanol and methyl formate, these motions allow a large number of transitions in the millimeter and sub-millimeter range.

The description of energy levels of methyl formate follows the standard nomenclature. For methanol, however, this is not the case, due to historical developments. These energy levels are labelled as J_k, where K can take on both positive and negative values as in Fig. 23. There is a similar scheme for naming energy levels of A type methanol, as $A_k^{\pm}$. A and E type methanol are analogous to ortho and para species, in that these states are not normally coupled by collisions.

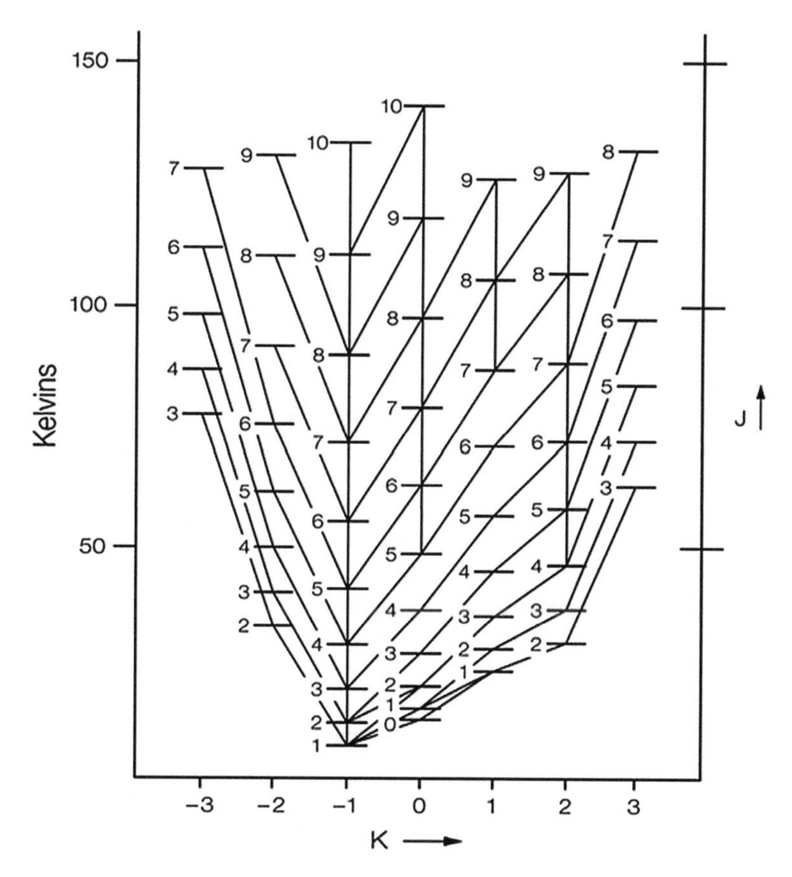

Fig. 23 Energy-level diagram of E type methanol, CH_3OH. This is an asymmetric top. The energy-level structure is typical of an almost prolate symmetric top molecule. The lines connecting the levels show the spontaneous transitions with the largest *A* coefficients from eac level; where the two largest *A* coefficients are within a factor of 2, both transitions are shown. Figure taken from Wilson et al. (2013) [52]

Torsionally excited states of methanol have been found in the interstellar medium. These are analogous to transitions from vibrationally excited states to the ground state of a molecule. Another complexity is that because of its structure, there are two dipole moments, along either the c or the a axis.

13 Astronomical Applications

In the following, methods to determine the parameters of molecular clouds are summarized.

13.1 Kinetic Temperatures

A crucial input parameter for the LVG calculations is $T_{\rm K}$. Linewidths of thermally excited species provide a definite value for $T_{\rm K}$, if the turbulent velocity can be neglected. The relation is

$$T_{\rm k} = 21.2 \left(m/m_{\rm H}\right) (\Delta V_{\rm t})^2 \ ,$$

where m is the molecule, m_H is the mass of hydrogen, and $\Delta V_{\rm t}$ is the FWHP thermal width. Such a relation has been applied to NH_3 lines (with success!) in quiescent dust clouds.

Historically, $T_{\rm K}$ was obtained at first from the peak intensities of rotational transitions of CO, which has a small spontaneous decay rate. From the ratio of CO to ^{13}CO line intensities, $\tau({\rm CO}) \gg 1$. After correcting for cloud size, from Eq. (174) $T_{\rm MB}$ and $T_{\rm ex}$ are directly related. In addition, the large optical depths reduce the critical density by a factor τ, the line optical depth, so that for the $J = 1 \rightarrow 0$ line the value of the critical density is $n^* \approx 50\ {\rm cm}^{-3}$. Then, the level populations are determined by collisions, so the excitation temperature for the $J = 1 \rightarrow 0$ transition is $T_{\rm K}$. Usually, it is assumed that the beam filling factor is unity; for distant clouds or external galaxies, the filling factor is clearly less.

An alternative to CO measurements makes use of the fact that radiative transitions between different K ladders in symmetric top molecules are forbidden. Then the populations of the different K ladders are determined by collisions. Thus a method of determining $T_{\rm K}$ is to use the ratio of populations in different K ladders of molecules such as NH_3 at 1.3 cm, or CH_3CCH and CH_3CN in the mm/sub-mm range. This is also approximately true for different K_a ladders of H_2CO. Since ratios are involved, beam filling factors play no role. Even for extended clouds, $T_{\rm K}$ values from CO and NH_3 may not agree. This is because NH_3 is more easily dissociated so must arise from the cloud interior. Thus for a cloud heated externally, the $T_{\rm K}$ from CO data will be larger than that from NH_3.

For NH_3, the rotational transitions, $(J+1, K) \rightarrow (J, K)$, occur in the far infrared and have Einstein A coefficients of order 1 s^{-1}; for inversion transitions, A values are $\sim 10^{-7}\ {\rm s}^{-1}$ (see e.g. Table 2). These non-metastable states require extremely high H_2 densities or intense far infrared fields to be populated. Thus NH_3 non-metastable states $(J > K)$ are not suitable for $T_{\rm K}$ determinations. Rather, one measures the inversion transitions in different metastable $(J = K)$ levels. Populations cannot be transferred from one metastable state to another via allowed radiative transitions. This occurs via collisions, so the relative populations of metastable levels are directly related to $T_{\rm K}$. The column densities are obtained from the inversion transitions from different metastable states using Eqs. (169)–(172). If the NH_3 lines are optically thick, one can use, in most cases, the ratios of satellite components to main quadrupole hyperfine components to determine optical depths. A large number of $T_{\rm K}$ determinations have been made using NH_3 in dark dust clouds. Usually these involve the $(J, K) = (1, 1)$ and (2,2) inversion transitions.

13.2 Linewidths, Radial Motions and Intensity Distributions

From the spectra themselves, the linewidths, $\Delta V_{1/2}$, and radial velocities, V_{lsr}, give an estimate of motions in the clouds. The $\Delta V_{1/2}$ values are a combination of thermal and turbulent motions. Observations show that the widths are supersonic in most cases. In cold dense cores, motions barely exceed Doppler thermal values. Detailed measurements of lines with moderate to large optical depths show that the shapes are nearly Gaussian. However, simple models in which unsaturated line shapes are Gaussians would give flat topped shapes at high optical depths. This is not found. More realistic models of clouds are those in which shapes are determined by the relative motion of a large number of small condensations, or clumps, which emit optically thick line radiation. If the motions of such small clumps are balanced by gravity, one can apply the virial theorem. Images of isolated sources can be used for comparing with models. One example is the attempt to characterize the kinetic temperature and the H_2 density distributions from spectral line or thermal dust emission data.

13.3 Determinations of H_2 Densities

If a level is populated by collisions with H_2, the main collision partner, a first approximation to the H_2 density can be achieved if one sets the collision rate, $n(H_2)\langle\sigma\, v\rangle$, equal to the A coefficient. The brackets indicate an average over velocities of H_2, which are assumed to be Boltzmann distributed. The H_2 density for a given transition which will bring T_{ex} midway between the radiation temperature and T_K is referred to as the *critical density*, and is denoted by n^*. That is $n^*\langle\sigma\, v\rangle = A$. However, this can be at best only an approximate estimate. For reliable determination of H_2 densities, $n(H_2)$, one must measure at least two spectral lines of a given species, estimates of the kinetic temperature, collision rates, and a radiative transport model. One can assume that the lines are optically thin, and use a statistical equilibrium model, but the present approach is to apply the LVG model [36]. Clearly, the more lines measured, the more reliable the result. For linear molecules such as CO or CS, it is not possible to separate kinetic temperature and density effects. For example, CO with $T_K = 10$ K will have a $J = 2 - 1$ line much weaker than the $J = 1 - 0$ line no matter how high the density. However, if the plot of normalized CO column density versus energy above the ground state shows a *turn over*, that is a decrease in intensity, it is possible to find a unique combination of T_K and $n(H_2)$.

13.4 Cloud Masses

13.4.1 Virial Masses

If we assume that only gravity is to be balanced by the motions in a cloud, then, for a uniform density cloud of radius R, in terms of the line of sight FWHP velocity, virial equilibrium requires:

$$\boxed{\frac{M}{M_{\odot}} = 250\left(\frac{\Delta v_{1/2}}{\mathrm{km\ s^{-1}}}\right)^{2}\left(\frac{R}{\mathrm{pc}}\right)}\ . \qquad (202)$$

Once again very optically thick lines should not be used in determining masses using Eq. (202).

Another probe of such regions is thermal emission from dust (Eq. 119)). Given the dust temperature, such data allow one to derive H_2 abundances and cloud masses.

13.4.2 Masses from Measurements of CO and ^{13}CO

CO is by far the most widespread molecule with easily measured transitions. The excitation of CO is close to LTE and the chemistry is thought to be "well understood", so measurements of CO and CO isotopomers (and dust emission measurements!) are the most important tool(s) for estimating masses of molecular clouds.

Even if all of the concepts presented in this section are valid, there may be uncertainties in the calculation of the column densities of CO. These arise from several sources which can be grouped under the general heading *non-LTE effects*. Perhaps most important is the uncertainty in the excitation temperature. While the ^{12}CO emission might be thermalized even at densities $< 100\ \mathrm{cm^{-3}}$, the less abundant isotopes may be sub-thermally excited, i.e., with populations characterized by $T_{\mathrm{ex}} < T_{\mathrm{K}}$ (this can be explained using the LVG approximation (139)). Alternatively, if the cloud in question has no small scale structure, ^{13}CO emission will arise primarily from the cloud interior, which may be either hotter or cooler than the surface; the optically thick ^{12}CO emission may only reflect conditions in the cloud surface. Another effect is that, although T_{ex} may describe the population of the $J = 0$ and $J = 1$ states well, it may not for $J > 1$. That is, the higher rotational levels might not be thermalized, because their larger Einstein A coefficients lead to a faster depopulation. This lack of information about the population of the upper states leads to an uncertainty in the partition function. Measurements of other transitions and use of LVG models allow better accuracy. For most cloud models, LTE gives overestimates of the true ^{13}CO column densities by factors from 1 to 4 depending on the properties of the model and of the position in the cloud. Thus a factor of two uncertainties should be expected when using LTE models. The final step to obtain the H_2 column density is to assume a ratio of CO to H_2. This is generally taken to be 10^{-4}. In spite of all these uncertainties, one most often attempts to relate measurements of the CO column density to that of H_2; estimates made using lines of CO (and isotopomers) are probably the best method to obtain the H_2 column density and mass of molecular clouds.

An LVG treatment of the dependence of the total column density on the line intensity of the $J = 2 \rightarrow 1$ line shows that a simple relation is valid for T_K from 15 K to 80 K, and $n(H_2)$ from $\sim 10^3$ to $\sim 10^6$ cm^{-3}. An assumption used in obtaining this relation is that the ratio of $C^{18}O$ to H_2 is 1.7×10^{-7}, which corresponds to $(C/H_2) = 10^4$, and $(^{16}O/^{18}O) = 500$. The latter ratio is obtained from isotopic studies for molecular clouds near the Sun. Then we have

$$N_{H_2} = 2.65 \times 10^{21} \int T_{MB}(C^{18}O, J = 2 \rightarrow 1)\, dv\,. \tag{203}$$

The units of v are km s^{-1}, $T_{MB}(C^{18}O, J = 2 \rightarrow 1)$ are Kelvin, and N_{H_2} are cm^{-2}. This result can be used to determine cloud masses, if the distance to the cloud is known, by a summation over the cloud, position by position, to obtain the total number of H_2.

The total cloud mass obtained from the methods above, or similar methods, is sometimes referred to as the 'CO mass'; this terminology can be misleading, but is frequently found in the literature.

13.4.3 Masses from the X Factor

In large scale surveys of the CO $J = 1 \rightarrow 0$ line in our galaxy and external galaxies, it has been found, on the basis of a comparison of CO with ^{13}CO maps, that the CO integrated line intensities measure mass, even though this line is optically thick. The line shapes and intensity ratios along different lines of sight are remarkably similar for both ^{12}CO and ^{13}CO line radiation. This can be explained if the total emission depends primarily on the number of clouds. If so, ^{12}CO line measurements can be used to obtain estimates of $N_{^{12}CO}$. Observationally, in the disk of our galaxy, the ratio of these two quantities varies remarkably little for different regions of the sky.

This empirical approach has been followed up by a theoretical analysis. The basic assumption is that the clouds are virial objects, with self-gravity balancing the motions. If these clouds are thought to consist of a large number of clumps, each with the same temperature, but sub-thermally excited (i.e., $T_{ex} < T_K$), then from an LVG analysis of the CO excitation, the peak intensity of the CO line will increase with $\sqrt{n(H_2)}$, and the linewidth will also increase by the same factor, as can be seen from Eq. (202). The exact relation between the integrated intensity of the CO $J = 1 \rightarrow 0$ line and the column density of H_2 must be determined empirically. Such a relation has also been applied to other galaxies and the center of our galaxy. However, the environment, including the Interstellar Radiation Field (ISRF), may be very different and this may have a large effect on the cloud properties. For the disk of our galaxy, a frequently used conversion factor is:

$$N_{H_2} = X \int T_{MB}(CO, J = 1 \to 0)\, dv$$
$$= 2.3 \times 10^{20} \int T_{MB}(CO, J = 1 \to 0)\, dv\,, \qquad (204)$$

where $X = 2.3 \times 10^{20}$ and N_{H_2} is in units of cm^{-2}. By summing the intensities over the cloud, the mass in $M_\odot$ is obtained. Strictly speaking, this relation is only valid for whole clouds. The exact value of the conversion factor, X, between CO integrated line intensity and mass of H_2 is a matter of some dispute.

13.5 Additional Topics

13.5.1 Freezing Out on Grain Surfaces

For H_2 densities $>10^6$ cm^{-3}, one might expect a freezing out of molecules onto grains for cold, dense regions. From a simplified theory for H_2 densities $> 10^6$ cm^{-3}, the time for a molecule-grain collision is 3000 years, short compared to other time scales. Then CO and most other molecules might be condensed out of the gas phase, so that spectroscopic measurements cannot be used as a probe of very dense regions. Empirically N_2H^+, NH_3 and H_2D^+ appear to remain in the gas phase even at low kinetic temperatures and high densities.

13.5.2 Self-shielding of CO

As is well established, CO is dissociated by line radiation. Since the optical depth of CO is large, this isotopomer will be self-shielded. If there is no fine spatial structure, selective dissociation will cause the extent of ^{12}CO to be greater than that of ^{13}CO, which will be greater than the extent of $C^{18}O$.

References

1. Altenhoff, W. J., et al.: Veroeff. Sternwarte Bonn **59** (1960)
2. Baars, J.W.M.: The Parabolic Reflector Antenna in Radio Astronomy and Communication, Astrophysics Space Science Library (2007)
3. Balser, D.S., et al.: Ap. J. **430**, 667 (1994)
4. Bachiller, R., Cernicharo, J.: Science with the Atacama Large Millimeter Array: A New Era for Astrophysics. Springer, Berlin (2008)
5. Baker, A. J., et al.: From Z Machines to ALMA: (Sub)millimeter Spectroscopy of Galaxies. ASP Conf. Ser. **75** (2007)
6. Bracewell, R.N.: The Fourier Transform and its Application, 2nd edn. McGraw Hill, New York (1986)
7. Cornwell, T., Fomalont, E.B.: Self calibration in synthesis imaging in radio astronomy. ASP Conf. Ser. **6**, 185 (1989)
8. DeLucia, F.J., et al.: Phys. Chem. Ref. Data **3**, 21 (1972)

9. Dicke, R.H.: Rev. Sci. Instrum. **17**, 268 (1946)
10. Downes, D.: Radio telescopes: basic concepts in diffraction-limited imaging with very large telescopes. NATO ASI Ser. **274**, 53 (1989)
11. Dutrey, A.: IRAM Millimeter Interferometry Summer School 2 (2000)
12. Gerlich, D., et al.: Phil. Trans. R. Soc. **364**, 3007 (2006)
13. Goldsmith, P.F.: Instrumentation and Techniques for Radio Astronomy. IEEE Press, New. York (1988)
14. Goldsmith, P.F.: Quasioptical Systems: Gaussian Beam Quasioptical Propagation and Applications. Wiley-IEEE Press, New York (1994)
15. Gordon, M.A., Sorochenko, R.L.: Radio Recombination Lines, their Physics and Astronomical Applications, Astrophysics and Space Science Library, vol. 282. Kluwer (2002)
16. Gull, S.F., Daniell, G.J.: Nature **272**, 686 (1978)
17. Gurvits, L., et al.: Radio Astronomy from Karl Jansky to Microjansky, EDP Sciences (2005)
18. Herbst, E.: Chem. Soc. Rev. **30**, 168 (2001)
19. Hildebrand, R.: Q. J. R. Astron. Soc. **24**, 267 (1983)
20. Högbom, J.: Astron. Astrophys. Suppl. **15**, 417 (1974)
21. Holland, W.S., et al.: M.N.R.A.S. **303**, 659 (1999)
22. Jackson, J.M., et al.: Ap. J. Suppl. **163**, 145 (2006)
23. Jenkins, F.A., White, H.E.: Fundamentals of Optics, 4th edn. McGraw-Hill, New York (2001)
24. Johnstone, D., et al.: Ap. J. Suppl. **131**, 505 (2000)
25. Kroto, H.W.: Molecular Rotation Spectra. Dover, New York (1992)
26. Krügel, E.: The Physics of Interstellar Dust Institute of Physics Press, Bristol (2002)
27. Love, A.W. (ed.): Electromagnetic Horn Antennas. IEEE Press, New York (1976)
28. Mangum, J.G., Wootten, A.: Ap. J. Suppl. **89**, 123 (1993)
29. Marrone, D.P., et al.: Ap. J. **654**, L57 (2007)
30. Martin-Pintado, J., et al.: Astron. Astrophys. **286**, 890 (1994)
31. Mezger, P.G., et al.: Astron. Astrophys. **228**, 95 (1990)
32. Melia, F.: The Galactic Supermassive Black Hole. Princeton University Press, Princeton (2007)
33. Motte, F., et al.: Astron. Astrophys. **476**, 1243 (2006)
34. Pawsey, J.L., Bracewell, R.N.: Radio Astronomy. Oxford University Press, Oxford (1954)
35. Reipurth, B., et al.: Protostars and Planets V. University of Arizona Press, Tucson (2007)
36. Rohlfs, K., Wilson, T.L.: Tools of Radio Astronomy. Springer, Berlin (2004)
37. Rieke, G.H.: Detection of Light: From Ultraviolet to the Submillimeter, 2nd edn. Cambridge University Press, Cambridge (2002)
38. Rybicki, G.B., Lightman, A.P.: Radiative Processes in Astrophysics. Wiley, New York (1979)
39. Sandell, G.: MNRAS **271**, 75 (1994)
40. Scaife, A.M.M., et al.: MNRAS **403**, L46 (2010)
41. Scheuer, P.A.G.: Radiation in Plasma Astrophysics. Proc. Int. Sch. Phys. Enrico Fermi **39**, 39 (1967)
42. Solomon, P.M., vanden Bout, P. A.: Ann. Rev. Astron. Astrophys. **43**, 677 (2005)
43. Sparke, L., Gallagher, J.S.: Galaxies in the Universe: An Introduction. Cambridge University Press, Cambridge (2000)
44. Stahler, S.W., Palla, F.: The Formation of Stars. Wiley-VCH, New York (2005)
45. Tielens, A.G.G.M.: The Physics and Chemistry of the Interstellar Medium. Cambridge University Press, Cambridge (2005)
46. Thompson, A.R., et al.: Interferometry and Synthesis in Radio Astronomy, 2nd edn. Wiley, New York (2001)
47. Thum, C., et al.: Publ. Astron. Soc. Pacific **120**, 777 (2008)
48. Townes, C.H., Schawlow, A.H.: Microwave Spectroscopy. Dover, New York (1975)
49. Vastel, C., et al.: Ap. J. **645**, 1198 (2006)
50. Wilson, T.L., Huettemeister, S.: Tools of Radio Astronomy: Problems and Solutions. Springer, Berlin (2005)
51. Wilson, T.L., et al.: Astron. Astrophys. **460**, 533 (2006)
52. Wilson, T.L., et al.: Tools of Radio Astronomy. Springer, Berlin (2013)

Star Formation with ALMA

S. Guilloteau

Contents

S. Guilloteau (✉)
Université Bordeaux 1, 33615 Pessac, France
e-mail: stephane.guilloteau@u-bordeaux.fr

S. Guilloteau
Laboratoire d'Astrophysique de Bordeaux (LAB), 33615 Pessac, France

S. Guilloteau
CNRS/INSU - UMR5804, 33615 Pessac, France

S. Guilloteau
bat B18N, CS 50023, 33615 Pessac, France

© Springer-Verlag GmbH Germany, part of Springer Nature 2018

M. Dessauges-Zavadsky and D. Pfenniger (eds.), *Millimeter Astronomy*,
Saas-Fee Advanced Course 38, https://doi.org/10.1007/978-3-662-57546-8_2

1 Introduction

Stars are believed to form from interstellar material through the gravitational collapse of dusty clouds. Interstellar medium is a very dynamical environment in which clouds of atomic gas (and its associated dust counterpart) form in warm medium fragments, perhaps as a result of turbulence or the passage of shock, and subsequently cool down and condense. Although the dust is a tiny fraction (of order 1%) of the total material mass, it plays a major role in the cloud evolution because its opacity can shield the cloud center from the interstellar UV field and dust surfaces act as a catalyst on which molecular hydrogen can form. For high enough column density, the combined effect of dust shielding and self-shielding of H_2 turn the initially predominantly atomic gas into molecular form. H_2 forms first, but more and more complex species such as CO, CN, and HCN, form during cloud evolution.

Occasionally, these so-called molecular clouds become gravitationally unstable and a global collapse begins that may eventually lead to star and planetary system formation. In this chapter, I provide insights into how the tools of millimeter and submillimeter astronomy can be used to unveil the details of this formation process. I emphasize the observational aspects and data analysis as opposed to the details of the physical processes, which are already treated in a number of other works.

1.1 Why mm/sub-mm Astronomy for Star and Planet Formation?

OPACITY

To understand why mm/sub-mm astronomy is essential, it is best to start with an optical picture of what is supposed to be a typical prestellar core. Figure 1a shows an optical image of the Bok globule B68 [81]. The dense cloud lies in front of the stellar background and is seen in silhouette because of the large extinction caused by the dust inside the core. Counting the stars as a function of distance from cloud center can reveal the extinction profile; however, the cloud opacity at optical wavelengths

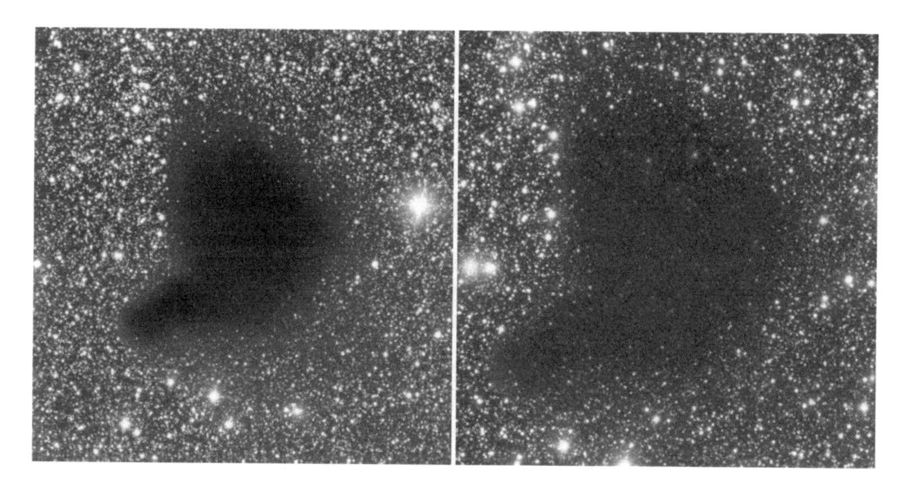

Fig. 1 An optical (left) and IR (right) image of the Bok globule B68

severely limits the depth to which such a study is sensitive. At near-infrared (NIR) wavelengths, the dust opacity is slightly lower and more details appear (Fig. 1b). A combined study of the extinction and reddening, using dust properties appropriate for the interstellar medium [88] facilitates the retrieval of the column density profile. However, the core remains so opaque that very little information can be recovered from these wavelengths.

If this cloud was homogeneously collapsing, the column densities would increase even further and all the interesting effects happening at the cloud center would remain hidden from our view. Moving to longer wavelengths is thus mandatory to unveil the physics of the densest, most opaque regions. After the NIR, the next good atmospheric window is the mm/sub-mm domain.

TEMPERATURES

Absorption studies require a bright background, which is not always available. Studying the emission is a more general tool: The same dust that absorbs the stellar light in Fig. 1 re-emits the same power at a longer wavelength. However, in these dense, star-free cores, the interstellar radiation field is efficiently blocked by the surrounding thick dust cloud and the dust temperature remains very low. The expected equilibrium temperature is around 7 K. The peak emission frequency for a blackbody at this temperature is 145 GHz.

MOLECULES

The last argument in favor of the mm/sub-mm domain is the molecular content of these clouds. The most abundant constituent, H_2, has no dipole moment and its lowest quadrupolar transition near 28 μm requires high excitation and is hardly observable from the ground. Helium also has no suitable transition. The next most abundant constituent is CO, a linear molecule with a rotation constant of 57 GHz. Many other molecules also exist in these molecular cores. However, since the rotation constant scale as the inverse of the molecule mass, all the simplest (diatomic) molecules, except hydrides like LiH, have their lowest transitions in the mm domain. In short, a combination of opacities, temperatures, and molecular constants make the mm/sub-mm domain ideal for the study of the earliest phases of star formation. We should consider ourselves lucky astronomers because even our atmosphere conspires positively with this favorable convergence by offering a transparency window right in this domain.

2 Dust as a Probe of Star Formation

From the introduction, it is clear that observations of dust emission at mm wavelengths is a key element in recovering the physical conditions in star-forming clouds. This chapter introduces the principle dust properties in this respect.

The dust can be characterized by three major aspects: composition, temperature, and size distribution. A fourth aspect, alignment, affects the polarization properties of the emitted radiation.

2.1 Size and Wavelength Dependence

Composition and size determines the absorption and scattering properties. A basic understanding comes from Mie theory, which allows us to derive the scattering and absorption properties of spheres. These properties are best explained using the absorption and scattering efficiencies $Q_{\rm abs}$ and $Q_{\rm sca}$, i.e., the ratios of the absorption (resp. scattering) cross sections to the geometrical cross section; we look at their dependencies as a function of the dimensionless parameter $x = 2\pi a/\lambda$, where a is the sphere radius and λ the wavelength.

For large, optically thick spheres, $x \rightarrow \infty$ and $Q_{\rm abs}$ tends toward 1.0; the sphere effective cross section is its geometrical cross section. The value of $Q_{\rm sca}$ also tends toward 1.0 as a result of the Babinet principle: The diffraction pattern of the sphere is that of an aperture with the same geometrical cross section and, thus, there is as much diffracted energy as energy intercepted by the sphere. This result holds for any sphere with a complex index of refraction whose value is not 1.0.

For small grains, $x \ll 1$,

$$Q_{\rm abs}(\nu) = 4\pi \left(\frac{2\pi a}{\lambda}\right) {\rm Im}\left[\frac{\epsilon(\nu)-1}{\epsilon(\nu)+2}\right], \tag{1}$$

and, provided $|\epsilon| x \ll 1$, the scattering efficiency is from Rayleigh theory

$$Q_{\rm sca}(\nu) = 4\pi \left(\frac{2\pi a}{\lambda}\right)^4 \left|\frac{\epsilon(\nu)-1}{\epsilon(\nu)+2}\right|^2, \tag{2}$$

where $\epsilon(\nu) = m^2$ is the complex dielectric constant (and m the refractive index). From Eqs. 1–2, the scattering is very inefficient at long wavelengths. For typical material, where $\mathrm{Im}(\epsilon) \ll \mathrm{Re}(\epsilon)$ and, thus, $d\mathrm{Re}(\epsilon)/d\nu = 0$ from the Kramers–Kronig relations, a first-order development gives

$$\frac{Q_{\mathrm{abs}}(\nu)}{a} = \frac{\nu}{c} \frac{12\,\mathrm{Im}(\epsilon(\nu))}{(\mathrm{Re}(\epsilon(\nu)) + 2)^2}. \tag{3}$$

Defining $\kappa(\nu)$ as the mass absorption coefficient, which for spheres of mean mass density ρ is

$$\kappa(\nu) = \frac{3}{4\rho} \frac{Q_{\mathrm{abs}}(\nu)}{a}, \tag{4}$$

we obtain

$$\kappa(\nu) = \frac{\nu}{c} \frac{9}{(\mathrm{Re}(\epsilon(\nu)) + 2)^2} \frac{\mathrm{Im}(\epsilon(\nu))}{\rho}, \tag{5}$$

showing that mass absorption coefficient is size independent in the limit of small grains. In the limit of large grains, as $Q_{abs} = 1$ (an equal amount of energy is scattered by diffraction at the edges), the mass absorption coefficient becomes wavelength independent, and scales with grains radius as $1/a$, i.e.,

$$\kappa(\nu) = \frac{3}{4\rho a}. \tag{6}$$

While the two asymptotic regimes are easily understood, for the intermediate grain size, $2\pi a \simeq \lambda$, interferences between the diffracted light and transmitted light result in resonances that can lead to large absorption efficiencies. The intensity of the resonances depend on the material properties. An example is given in Fig. 2, which shows κ as a function of grain radius for two different materials [80].

In the absence of detailed knowledge of the material properties, this mass absorption coefficient is often parameterized, *in the mm and submm domain only*, as

$$\kappa(\nu) = \kappa_0 \left(\frac{\nu}{\nu_0}\right)^{\beta}, \tag{7}$$

and one of the observer's issues is to use the most appropriate values for κ_0 and β. From the previous equations, one can see that $\beta \rightarrow 0$ for large grains (Eq. 5).

2.2 *Size Distribution*

In the diffuse interstellar medium, [88] showed that the extinction curve could be adequately reproduced by a mixture of graphite and silicate grains with a grain size

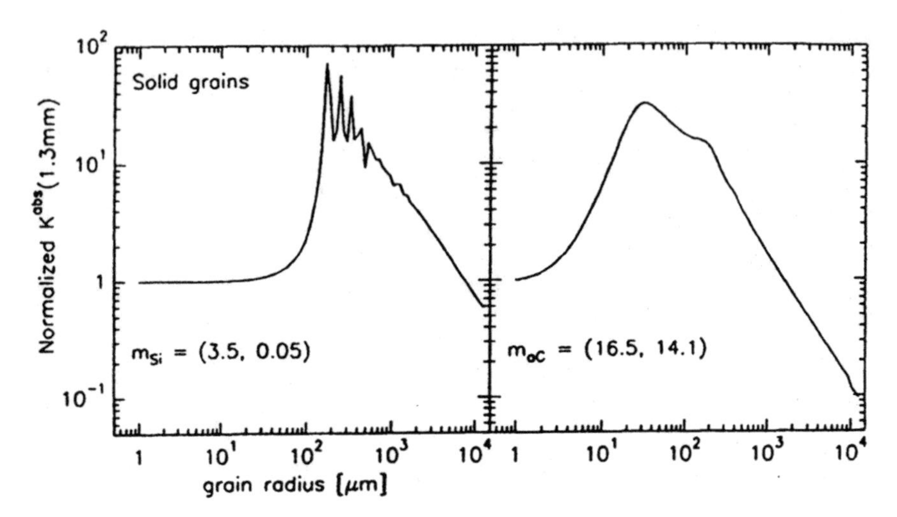

Fig. 2 Absorption coefficient at $\lambda = 1.3$ mm as a function of grain size for silicate and graphite spheres (from Kruegel and Siebenmorgen 1994 [80])

distribution following a power law described by $n(a) \propto a^{-\gamma}$ with $\gamma \simeq 3.5$ and a typical maximum grain size $a_+ \sim 0.1$ μm. Such a distribution can be obtained by a collisional cascade in which initially large grains are broken into small grains. What happens to the absorption coefficient $\kappa(\nu)$ when such size distributions are considered? This involves shifting the curves presented in Fig. 2 both in abscissa a from a_- to a_+ and ordinate to take the relative fraction of each size of grain into account, i.e., the exponent γ., We then integrate the results, i.e., convolve the curves in Fig. 2 by some function that depends on γ. A typical result is shown in Fig. 3. For small a_+, the value of κ_0 remains constant since all grains are much smaller than the wavelength. For large a_+, the absorption is dominated by large grains for small γ, which leads to $\kappa_0(a_+) \propto 1/a_+$ as the volume to surface ratio is $1/a$. For progressively larger γ, κ_0 flattens out and asymptotically reaches the same value as for small a_+, since small grains become dominant. In parallel, the frequency exponent β changes. For large a_+ and small γ, large grains dominate and, therefore, β tends toward zero. For $a_+ \approx \lambda_0$, both the absorption coefficient κ_0 and the exponent β exhibit a maximum. The maximum absorption coefficient κ_0 is only larger by a factor of a few compared to the small grain limit, however. This maximum "enhancement" depends on the strength of the resonances, i.e., on the material properties, and can increase to about 10–20 for grains that resemble those of the interstellar medium, but values around 2–3 are more likely with a wider range of materials. A useful approximate behavior was derived by [35]: when $a_+ > 3\lambda$, $\beta \approx (\gamma - 3)\beta_s$, where β_s is the small grain opacity index. The relation is valid for $3 < \gamma < 4$. This shows that grain growth, either by increasing a_+ or by flattening the number density distribution (reducing γ), results in a decrease of the opacity index β.

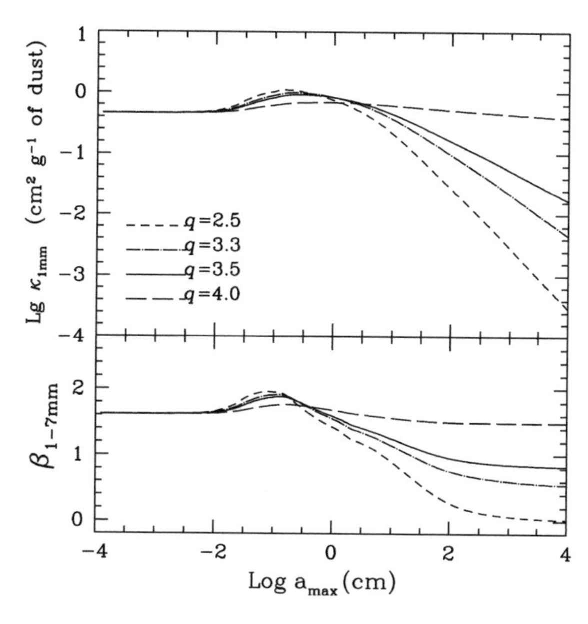

Fig. 3 Absorption coefficient at $\lambda = 1.3$ mm as a function of maximum grain size for a power-law size distribution (from Natta et al. 2004 [94])

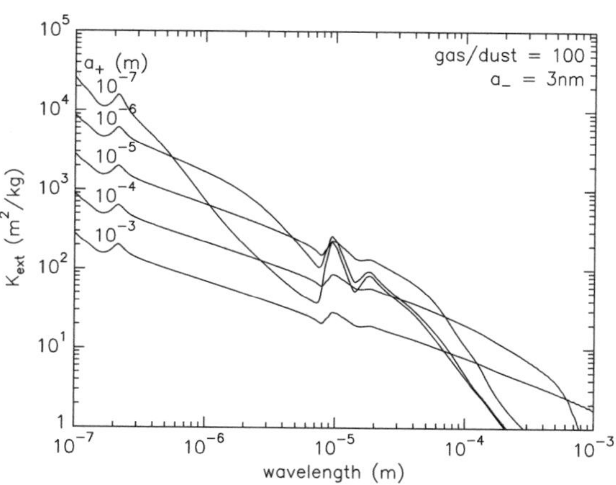

Fig. 4 Absorption coefficient as a function of wavelength for various size distributions

Figures 2 and 3 are useful to consider the effect of grain size on the mm continuum emission from dust, however, the grain size dependence also affects the UV extinction, which is another essential feature of the dust. Figure 4 shows the overall extinction curve, from the UV domain to the mm wavelengths, for different size distributions. At short wavelengths, the mass absorption coefficient decreases with a_+. In the case of $\gamma = 3.5$, it is easy to show that $A_V \propto 1/\sqrt{a_+ a_-}$. Table 1 gives some simple dependencies of the (short wavelength) mass absorption coefficient.

Table 1 Mass absorption coefficient scaling law

γ	κ
2.5	$\propto 1/a_+$
3.0	$\propto \log(a_+/a_-)/a_+$
3.5	$\propto 1/\sqrt{a_+a_-}$
4.5	$\propto 1/(a_- \log(a_+/a_-))$

2.3 *Dust Material Properties*

Although the size and wavelength dependencies are general features, the absolute value of the absorption coefficient depends on the detailed material properties. Materials such as silicate (glasses) or graphite have very different dielectric constants, making their mass absorption coefficient differ by factors of a few. The mass absorption coefficient of several materials is shown in Fig. 5. This figure also reveals a dependency of the spectral index β on both frequency and temperature. In general, the absorption coefficient decreases with temperature. Crystalline insulators lead to $\beta = 2$. Amorphous silicates show variations of β with temperatures between 1.6 at 300 K and 3 at 10 K [16].

The temperature dependence is mainly due to several physical effects, such as the existence of two level systems (TLS). The TLS are the result of vacancies in the (disordered) lattice and lead to tunneling between states of similar potential energy [1]. These processes depend on temperature and can be subdivided into different contributions: resonant absorption, tunneling relaxation, and hopping relaxation. Disordered charge distribution (DCD), which exists in amorphous material, leads to a temperature independent contribution. The DCD results in the existence of acoustic modes in the grains at frequencies $\nu > \pi V_t/a$, where V_t is the transverse sound velocity in the material. This critical frequency is around 90 GHz for 0.1 μm grain size. Hence, the temperature dependence of the absorption coefficient is most critical at long wavelengths (see Fig. 6). The importance of TLS and DCD depends more on the amount of disorder than on the material constitution [91].

The temperature dependence is often ignored when analyzing observations. [37] studied the sub-mm emission from several clouds with the balloon-borne sub-mm telescope PRONAOS, sampling a wide variety of conditions (cirrus clouds, star-forming regions, a galaxy, etc.), and thus a wide range in expected dust temperatures. The small telescope provided a relatively low angular resolution of a few arcminutes. However, the multiband capacity enabled the production of the spectral energy distribution (SED) in the sub-mm domain. A few examples are given in Fig. 7. The SED adjustment is made assuming optically thin emission at a fixed temperature, i.e., $I_\nu \propto B_\nu(\nu, T)\nu^\beta$.

This suggests that the spectral emissivity index β is a decreasing function of the temperature. This trend is confirmed with a larger set of independent measurements as shown in Fig. 8. Although the scatter is large, $\beta = 2.5/(1 + 0.2T_\mathrm{d})$ provides a good fit to the data. The index β itself is not sufficient to characterize the dust emissivity

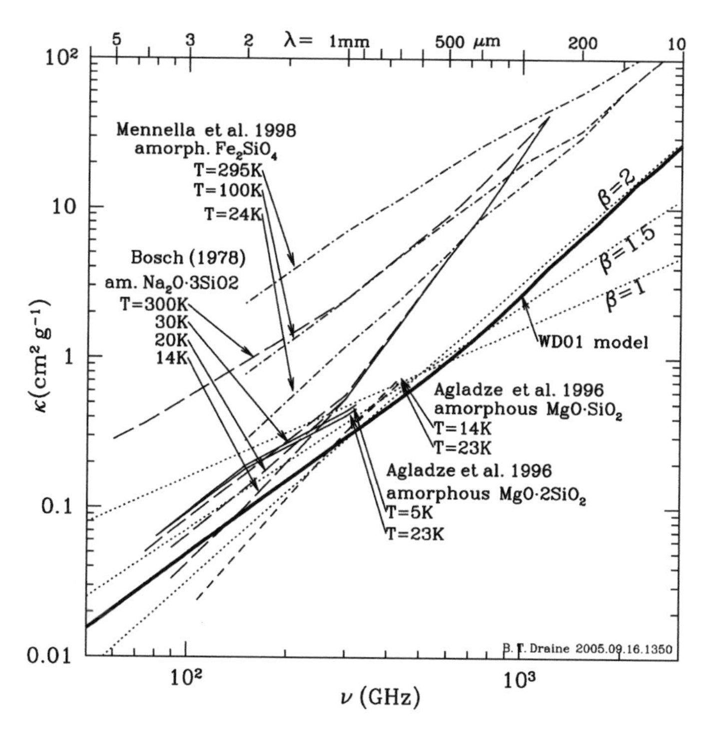

Fig. 5 Temperature dependence of the absorption coefficient as a function of wavelength (from Draine 2006 [35])

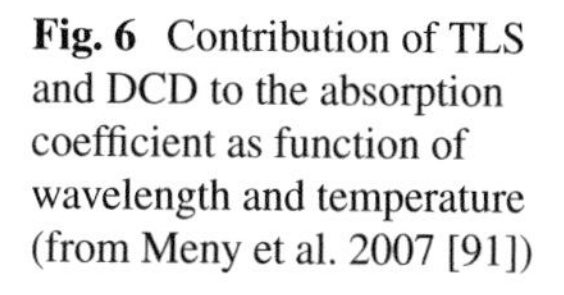
Fig. 6 Contribution of TLS and DCD to the absorption coefficient as function of wavelength and temperature (from Meny et al. 2007 [91])

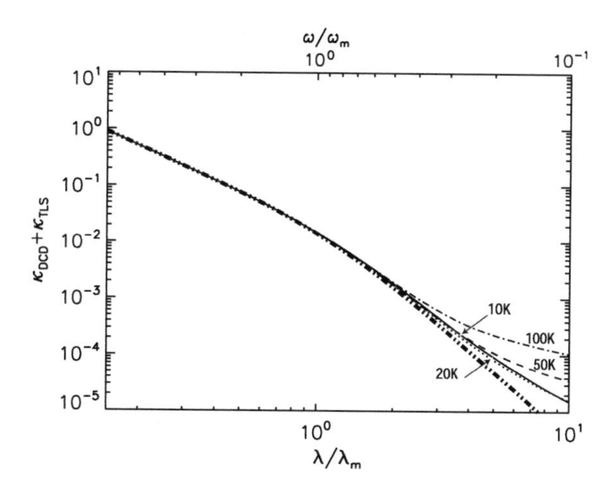

and its temperature dependence. A comparison with the results presented in Fig. 5 suggests that the emissivity near 1000 GHz (300 μm) may be relatively insensitive to temperature, while a strong decrease with temperature is seen at longer wavelengths.

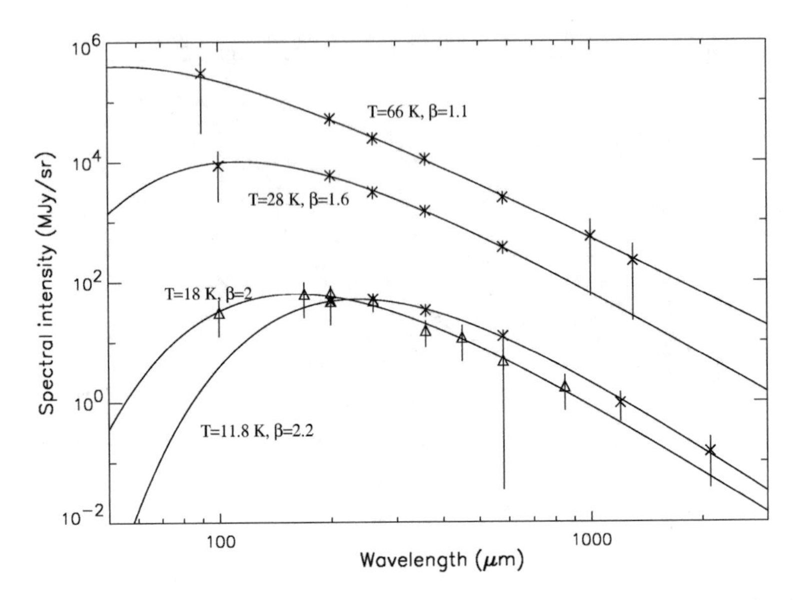

Fig. 7 Sub-mm SED from various regions. From top to bottom, the crosses represent data from the Orion Molecular Cloud 1, M 17 North, and Cloud 2 in Orion, and triangles represent data from the NGC 891 galaxy. The error bars are plotted within the 3 σ confidence level. The solid lines represent the results of the fit by a single modified blackbody with parameters T and β indicated (from Dupac et al. 2003 [37])

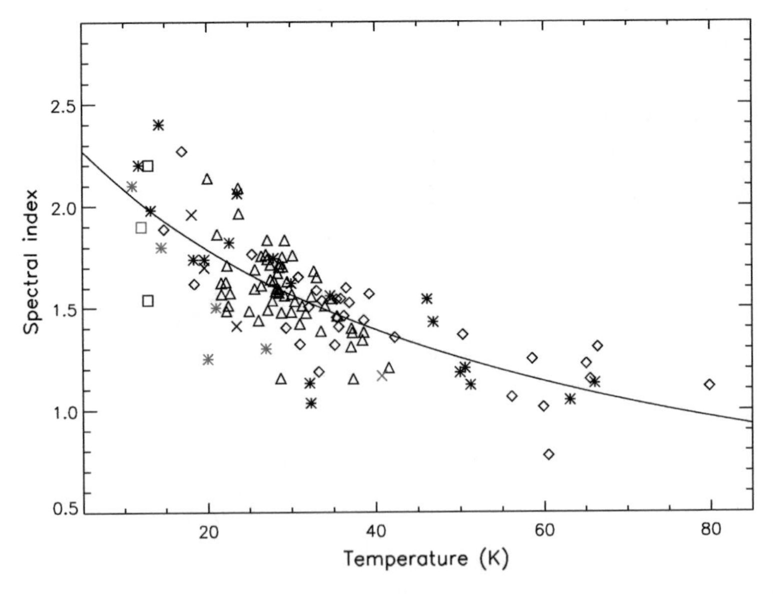

Fig. 8 Spectral index versus temperature for fully independent pixels in Orion (black asterisks), M 17 (diamonds), Cygnus (triangles), Ophiuchi (gray asterisks), Polaris (black squares), Taurus (gray square), NCS (gray cross), and NGC 891 (black crosses). The solid line is the result of the best hyperbolic fit (from Dupac et al. 2003 [37])

Much work remains to be done on the temperature dependence of the dust emissivity, especially in regions in which dust grains significantly differ from the average dust grains of the interstellar medium.

Fluffiness and Ice Coating

From Eqs. 1–2, the mass absorption coefficient scales as $1/\rho$, where ρ is the material density. We thus expect that, in general, fluffy grains will be more efficient absorbers than compact grains.

Another physical phenomenon playing a role in the dust emission properties at mm/sub-mm wavelengths is the presence of ice mantles on the grain surfaces. The relative importance of dust coagulation and ice mantles was studied in detail by [96] in the context of Mie theory. While grain growth (by coagulation, leading to fluffy grains) can lead to substantial increase in the mass absorption coefficient at mm wavelength, the presence of ice mantles reduces the overall effect of grain growth (see Fig. 9).

Recommendation for Astronomers

The dust absorption coefficient thus depends on four major effects: behavior at low temperature, grain growth, fluffiness, and the existence or inexistence of ice mantles. These effects are pulling in different directions, but conspire to give an absorption coefficient of order 1 $cm^2\ g^{-1}$ (per gram of dust, not per gram of the associated gas) near 1.2 mm within a factor of $2 - 3$. This is only a factor 2–4 more than derived using standard interstellar dust properties. The uncertainty on the mass absorption coefficient is directly transmitted to the mass of material derived from the dust continuum.

A convenient parametrization in the mm domain is

$$\kappa(\nu) = \kappa_{1.2}(\nu/250\ \mathrm{GHz})^{\beta}, \tag{8}$$

where $\kappa_{1.2}$ is the absorption coefficient at 1.2 mm (250 GHz). Similar parametrizations with different pivot frequencies have been used, especially in the context of protoplanetary disks [9]. Correcting for the reference frequency is unfortunately often needed to properly compare results from several sources. Furthermore, it should be emphasized that the behavior at low temperature has remained poorly explored so far, and that at very low temperatures the dust emissivity used in many astrophysical studies may have been overestimated. In the above parametrization, choosing a reference frequency around 1 THz may allow us to handle most of the temperature dependence through variations on β only. It is, however, unclear whether the empirical fit $\beta_s(T_\mathrm{d}) = 2.5/(1 + 0.4T_\mathrm{d})$, derived by [37], can be applied to all sources with some proper scaling for grain growth, such as $\beta = (\gamma - 3)\beta_s$ derived by [35].

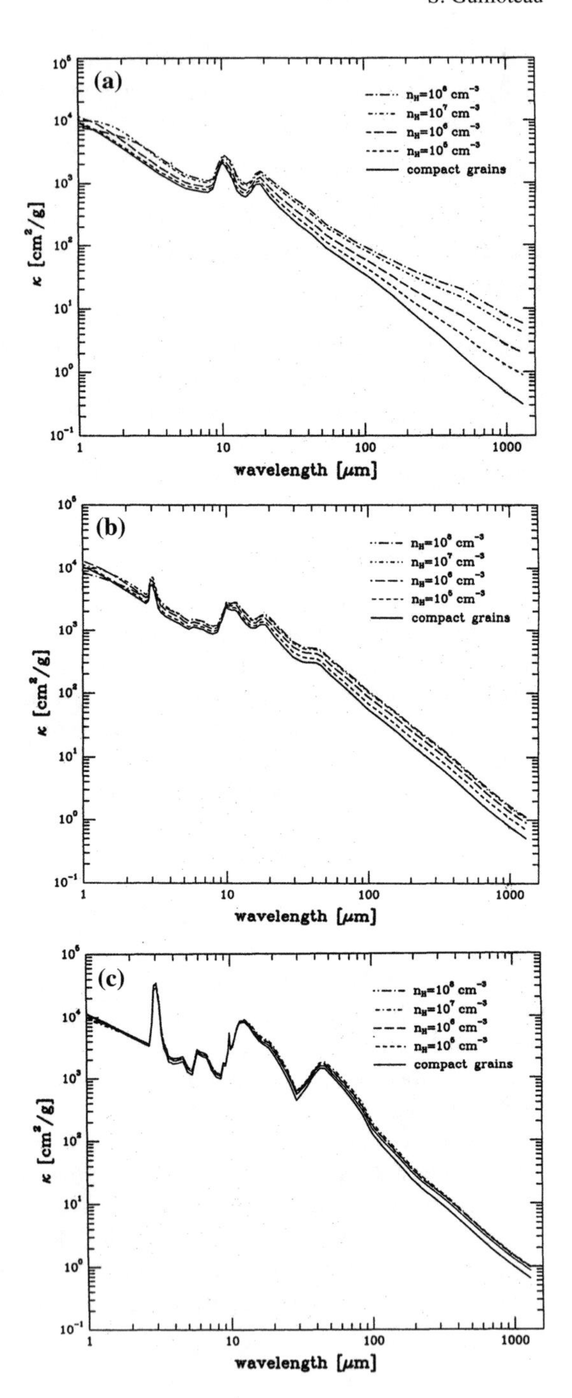

Fig. 9 Dust opacity for fluffy and ice-coated grains as function of wavelength (from Ossenkopf and Henning 1994 [96])

2.4 Dust as a Mass Tracer

In the optically thin regime, and under the Rayleigh-Jeans assumption, which is appropriate except perhaps at the lowest temperatures (<10 K) and highest frequencies (above 300 GHz, i.e., the sub-mm domain), the dust emission from a source leads to a flux density

$$S_\nu = \frac{2kT_{\rm dust}}{\lambda^2 D^2}\kappa_\lambda M, \tag{9}$$

where D is the source distance, $T_{\rm dust}$ the dust temperature, and κ_λ the dust emissivity per unit mass *of dust + gas* at wavelength λ and M the total mass. Plugging in appropriate numbers, with a gas-to-dust mass ratio of 100,

$$S_{250\ \rm GHz}(Jy) = 12\left(\frac{T}{15\ \rm K}\right)\left(\frac{100\ \rm pc}{D}\right)^2\left(\frac{M}{1\ \rm M_\odot}\right). \tag{10}$$

Wilson (Sect. 4.1, Eq. 49) shows that a telescope like the IRAM (Institut de Radio Astronomie Millimétrique) 30 m telescope equipped with a background limited bolometer has a typical point source sensitivity of about 40 mJy per second, or around 2.5 mJy in 10 min, down to a threshold of about 0.4 mJy below which the noise no longer integrates down as $1/\sqrt(t)$. In the imaging mode with the MAMBO-2 bolometer, the 30 m can image a $150 \times 150''$ with $11''$ resolution to about 3 mJy rms in two hours, when sky-noise filtering is efficient. These sensitivities, when compared with the expected flux from cloud cores, clearly show that molecular clouds and prestellar cores of even very low masses can be readily detected by current mm telescopes equipped with mm/sub-mm bolometers up to a few kpc. Unbiased large-scale imaging surveys are possible down to the brown dwarf mass limit for nearby clouds (e.g., ρ Oph or Taurus-Auriga region, 100–150 pc), where the linear resolution is even sufficient to resolve the individual condensation.

Furthermore, as there is a simple relation between the observed flux density and mass, this capacity has been used to derive the mass spectrum of clumps by, e.g., [93]. In these studies, the choice of the observing frequency is a compromise dictated by several considerations. Sensitivity, field of view, and angular resolution are the prime factors to consider. Higher frequencies yield more flux and are thus more sensitive to low-mass cores. However, beyond the mere detection, the flux-to-mass conversion may be affected by deviation from the Rayleigh-Jeans approximation, leading to a bias against cold cores at high frequencies. On the other hand, the evolution of dust emissivity with temperature (and grain size) is more important at the lowest frequencies.

3 Using Molecules

The mass distribution is only one basic aspect of the physics of star formation. Other very essential quantities are the temperature, which influences all emission processes; the chemical composition, which in part controls the cooling processes; and, of course, the kinematics, which traces the dynamical evolution. To study these, molecular observations are essential.

To understand how molecules can be used to study these aspects, it is useful to keep in mind some basic properties of the rotational spectra of relatively simple molecules. The detailed equations have been exposed in T. Wilson lectures.

Line intensities scale as ν^3, which means high J lines should be stronger. Line intensities also scale as μ^2, hence, molecules with large dipole moment are easier to detect. However, radiative decay competes with collisional excitation, so molecules with large μ are more difficult to thermalize (high J lines).

Determining which molecule/transition pair is optimal (or at least suitable) to study a given problem is an ongoing problem when trying to observe the star formation process. I present here, in a very simple way, a few useful relations and criteria in this respect.

3.1 Useful Simple Relations

Brightness of the $J = 1 - 0$ Transition

The $J = 1 - 0$ transition of a linear molecule has a very interesting property in the optically thin regime

$$T_{\rm b} = (1 - \exp(-\tau))T_{\rm ex}, \tag{11}$$

where

$$\tau \propto (N_u - N_l)/\Delta V \tag{12}$$

is the line opacity

$$\tau \propto \left(1 - \exp(-\frac{h\nu}{kT_{\rm ex}})\right) \frac{N_{mol}}{Z\Delta V}. \tag{13}$$

Assuming the rotational ladder is reasonably thermalized, $T_{\rm ex} = T_k$, and that the temperature is high enough, $h\nu << kT_k$, the partition function is $Z \propto 1/T_k$. Hence,

$$T_{\rm b} \propto 1/T_k. \tag{14}$$

Transition of Maximum Opacity

Using $a = hB/kT$, and provided $T > hB/2K$, the partition function is $Z = kT/hB = a^{-1}$. Hence, the line opacity of the $J \rightarrow J - 1$ transition

$$\tau_J = \frac{8\pi^3\mu^2}{3hZ}\frac{N\rho}{\Delta V}J(e^{2aJ}-1)e^{-aJ(J+1)} \tag{15}$$

is maximum for $J_m = a^{-1/2}$, at frequency $\nu_m = 2Ba^{-1/2}$. Numerically,

$$J_m = 4.6\sqrt{T(\mathrm{K})/B(\mathrm{GHz})} \quad \text{and} \quad \nu_m(\mathrm{GHz})) = 9.13\sqrt{T(\mathrm{K})B(\mathrm{GHz})}. \tag{16}$$

Some knowledge of the kinetic temperature, as well as the list of rotation constants of the most simple and abundant molecules, readily indicates which transitions are the strongest and at which frequencies they should be searched. With $T = 10$ K, as in cold cores, $B = 50$ GHz, as for the simplest molecules (CO, CN, HCN, etc.), $\nu_m \simeq 225$ GHz. More complex molecules require somewhat lower frequencies.

Brightness of the Maximum Opacity Transition

Under the same assumption of high temperature,

$$\tau_J = \frac{8\pi^3\mu^2}{3hZ}\frac{N\rho}{\Delta V}J(e^{2aJ}-1)e^{-aJ(J+1)}, \tag{17}$$

$$\Delta T_\mathrm{b} = J_\nu(T)\tau_J = \frac{8\pi^3\mu^2 B}{3k}\frac{N\rho}{\Delta V}J(e^{2aJ}-1)e^{-aJ(J+1)}. \tag{18}$$

We recover the previously demonstrated fact for $J = 1-0$, i.e.,

$$\Delta T_\mathrm{b} = \frac{8\pi^3\mu^2 B}{3k}\frac{N\rho}{\Delta V}(1-e^{-2a}) = \frac{8\pi^3\mu^2 B}{3k}\frac{N\rho}{\Delta V}\frac{2hB}{kT} \propto \frac{N}{T\Delta V}, \tag{19}$$

but for J_m transition

$$\Delta T_\mathrm{b} = \frac{8\pi^3\mu^2 B}{3k}\frac{N\rho}{\Delta V}\frac{1}{\sqrt{a}}(e^{2\sqrt{a}}-1)e^{-(1+\sqrt{a})} \approx \frac{8\pi^3\mu^2 B}{3k}\frac{N\rho}{\Delta V}0.8 \propto \frac{N}{\Delta V}. \tag{20}$$

In other words, the J_m transition's brightness is essentially temperature independent provided that it remains optically thin. In addition, the opacity ratio, $\tau(J_m)/\tau(1) = 1/(ea)$.

4 Prestellar Cores

4.1 Using Dust

The beam size (10–15") of large telescopes equipped with bolometers provides an angular resolution of 1000 AU at the distance of nearby star-forming regions (100–150 pc), which is sufficient to resolve prestellar cores. Thus, continuum observations may be a good tool to study the distribution of mass in prestellar cores. However, the brightness sensitivity and, as a consequence, the sensitivity to column density of material, is too low to follow the dust distribution to radii beyond a few 1000 AU. Under the assumption that the cores have spherical symmetry, using radial averaging allows us to reach typical radii around 20,000 AU. Figure 10 shows an example of radial emissivity profiles for several dense cores [118]. The line is a fit of the expected intensity profile for densities varying as $n(r) = n_0/(1 + (r/r_0)^\alpha)$ and assuming a uniform temperature. In deriving this theoretical emission profile, the overall imaging process resulting from the bolometer observations must be accounted for as beam convolution, on-the-fly(OTF) scanning, and wobbling that distort the sky image [92].

The derived profiles are very similar to those derived from visual extinction measurements in B68 by [81]. These profiles correspond closely to the family of Bonnor–Ebert spheres. Bonnor–Ebert spheres are spheres of gas in pressure equilibrium under self-gravity. The basic equations are

$$\frac{dP(r)}{dr} = -\frac{GM(r)\rho(r)}{r^2} \quad \frac{dM(r)}{dr} = 4\pi r^2 \rho(r),$$

and assuming isothermal conditions

$$P(r) = a^2 \rho(r) = \frac{kT}{\mu m_H} \rho(r).$$

Using $D = \rho/\rho_c, \xi = (r/a)(4\pi G\rho_c)^{1/2}$ and $D = e^{-\phi(\xi)}$, this can be transformed into the Lane-Emden differential equation

$$\frac{d^2\phi}{d\xi^2} + \frac{2}{\xi}\frac{d\phi}{d\xi} = e^{-\phi(\xi)}$$

under the boundary constraint

$$\phi(0) = \frac{d\phi}{d\xi}(0) = 0.$$

The solution only depends on $\xi_{\max} = \xi(r_{\max})$ and the sphere is stable if $\xi_{\max} < 6.451$, which corresponds to $dP/dV(r_{\max})$ being positive.

The emission profiles that are determined in the mm/sub-mm thus appear to closely match the extinction studies made in the visible or NIR. However, the extinction studies require a dense enough stellar background to provide sufficient angular resolution and can be confused by foreground stars. Thus, these extinction studies are limited to nearby clouds. On the other hand, mm/sub-mm continuum emission can be used for any distance, provided the angular resolution and brightness sensitivity of the telescope remain sufficient. The largest telescopes provide an angular resolution ($10 - 15''$) that unfortunately limits these types of studies to distances around

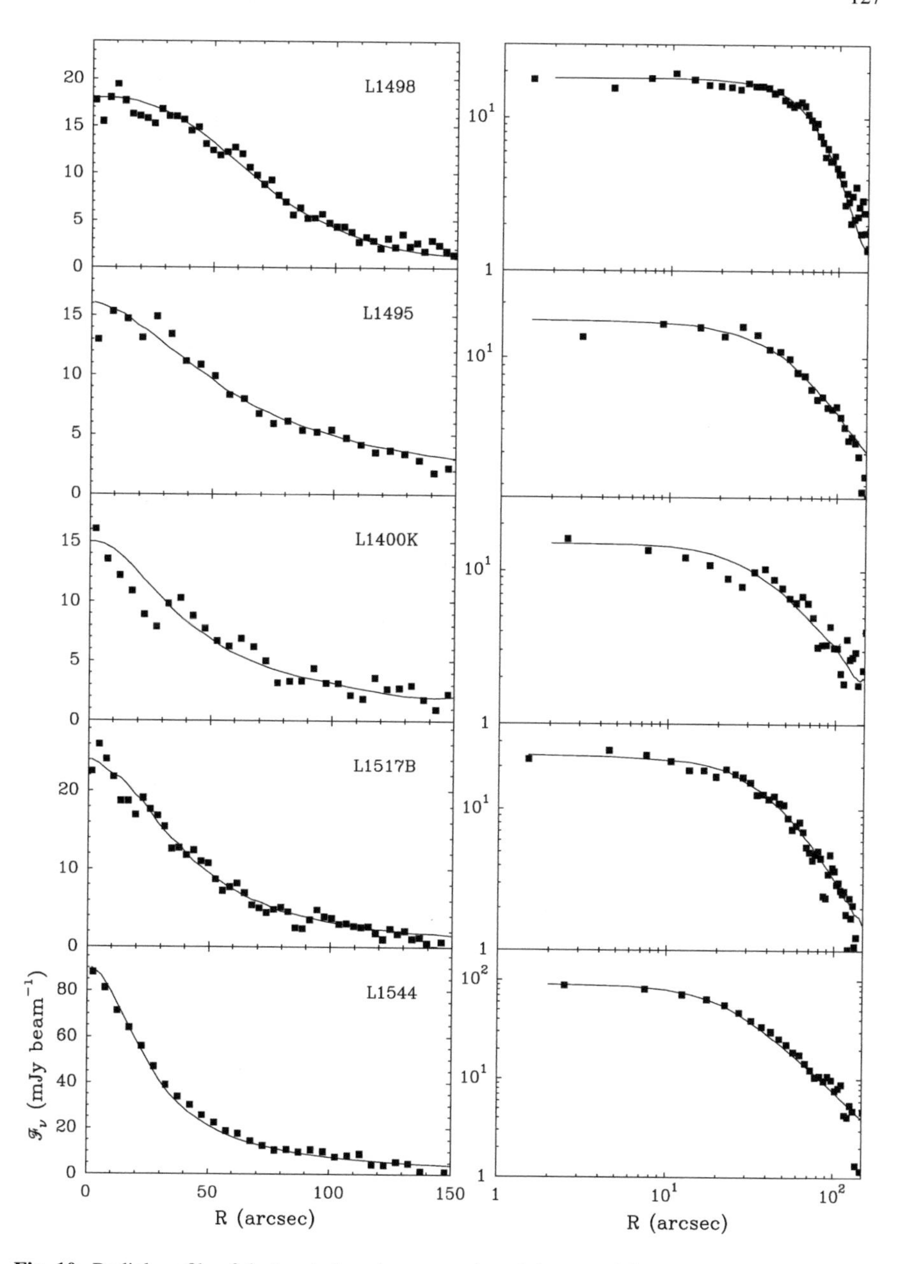

Fig. 10 Radial profile of dust emission at mm wavelength in several dense cores (from Tafalla et al. 2002 [118])

200 pc. Interferometers can provide much better resolution, but in this case the limit is the surface brightness sensitivity.

Another limitation of the use of continuum emission to probe the density structure is its sensitivity to the dust temperature. In the above study, the dust temperature was assumed to be constant. This is unlikely to be true, however, and changing the temperature changes the link between brightness and dust column density. This is illustrated by [79], who studied the ratio of dust emission over visual extinction by comparing a bolometer image of the IC5146 dark cloud with an extinction map derived from star counts. The flux-to-A_V ratio can either be interpreted as an indication of varying dust opacity with distance to the cloud core or as a temperature gradient with lower temperatures inside the cloud than at the edges (see Fig. 11).

Temperature gradients are expected in prestellar cores, as these cores are heated from the outside and have sufficient extinction to prevent the interstellar radiation field (ISRF) from reaching the innermost regions. The dust temperature is largely controlled by the radiation field. Gas-grain collisions only play a limited role and when gas-grain collisions are important, dust imposes its temperature on the gas. In general, cosmic ray heating, which is an indirect process that results from secondary UV photons along with collisions with the heated gas, also has a limited impact even on the most obscure regions. Since the ISRF is attenuated by extinction, dust in cloud cores are expected to be colder than at the cloud edges. The ISRF depends on the environment. For cores near giant molecular clouds, the infrared field dominates the energy balance because the UV field is very quickly stopped by the outer edges of the cores (see [45]). In this case, as the IR dominates the energy budget, the temperature of grains smaller than $1-2\ \mu$m is independent of their size. However, this temperature is not independent of their composition since it depends on the long wavelength grain emissivity. Larger grains may have size dependent temperatures and are colder than the small grains.

A detailed study of the impact of dust temperature on the interpretation of radial emission profiles of prestellar cores has been performed by [43]. They show that the dust temperature is expected to be 14–16 K at the cloud edges, as most authors assumed, but can drop down to about 7–8 K in the core center. At these temperatures, the Rayleigh-Jeans approximation is no longer accurate and, hence, the emission critically depends on the temperature structure. As the temperature is lower in the core, it contributes to a flattening of the emission. Accordingly, the density profile is more centrally peaked than a constant temperature analysis predicts.

In the pressure equilibrium, the temperature has a feedback on the density profile

$$\frac{dP(r)}{dr} = k\left(n\frac{dT}{dr} + T\frac{dn}{dr}\right) \tag{21}$$

$$\frac{dn}{dr} = -\left(\frac{\mu m_H G}{k}\right)\frac{M(r)n}{Tr^2} - \frac{n}{T}\frac{dT}{dr}. \tag{22}$$

The negative temperature gradient results in a smaller core with a steeper density gradient than in the isothermal case. At any radius, there is thus less mass enclosed,

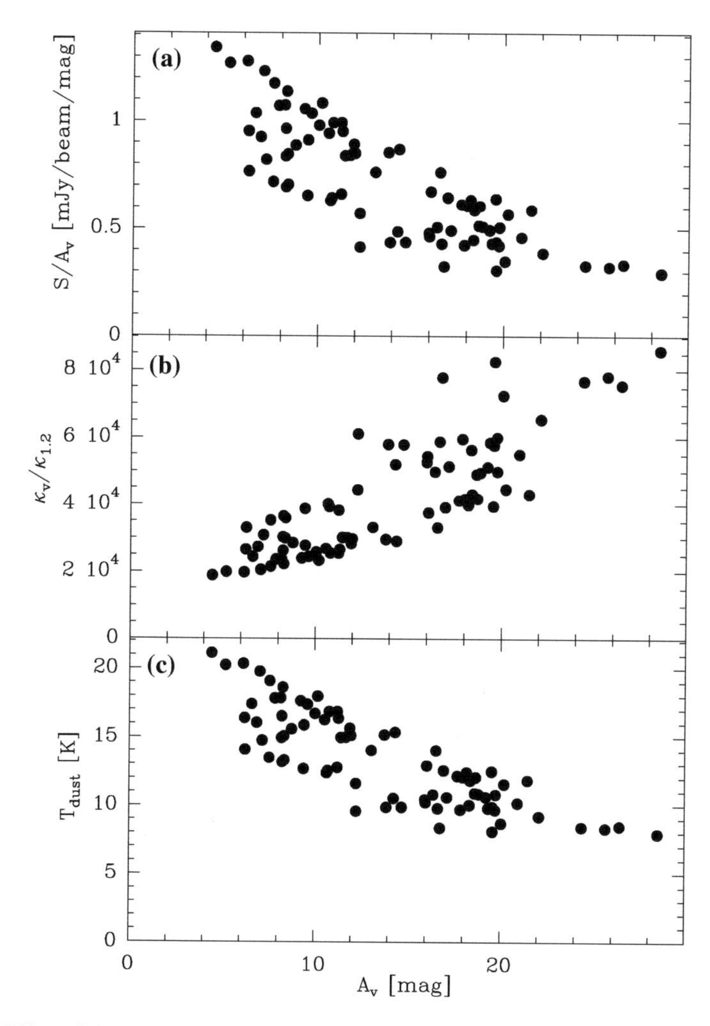

Fig. 11 Effect of the temperature gradient on the flux to A_v ratio in IC5146 (from Kramer et al. 1998 [79])

so the density profile outside is shallower. A family of solutions can be computed iteratively, starting from the isothermal case. The temperature structure is then derived via radiative transfer to solve for the dust temperature as a function of radius. This temperature is then used to solve the pressure equilibrium Eqs. 21–22. The process usually converges in two such iterations. The emission profile can then be computed, including the effects of beam switching and convolution by the telescope beam. Strictly speaking, one should use the gas temperature in deriving the pressure equilibrium. The gas temperature is a complex issue, and there are many heating and cooling processes involved. The gas can be heated by chemical processes, i.e., exothermic reactions, such as the formation of H_2, the photoelectric effect, gas-grain

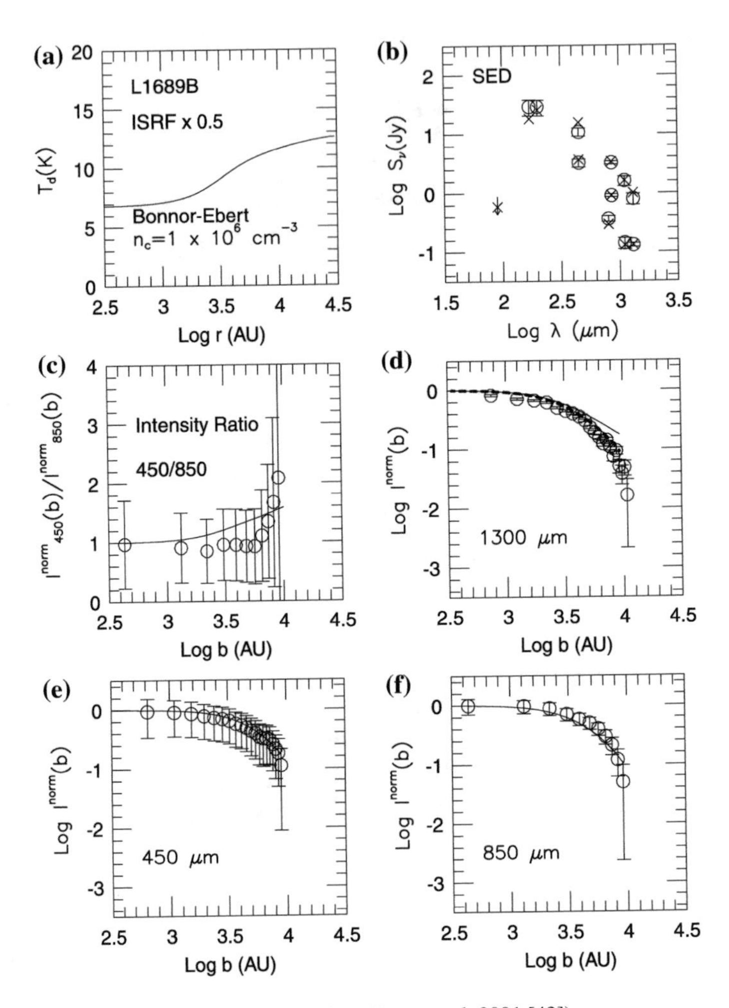

Fig. 12 Density and temperature profiles (from Evans et al. 2001 [43])

collisions (which can be a heating or a cooling mechanism), or cosmic rays. The cooling is mostly controlled by the radiation through rotational lines of the most abundant molecules. Accordingly, the kinetic temperature is expected to depend on the chemical composition of the medium, which itself depends on temperature and density. Solving this complex problem correctly is challenging, and one often relies on an approximate solution. Fortunately, at the high extinction prevailing in the cores, and given the high densities in the cores, gas-grain collisions are expected to be the dominant effect and one expects the dust temperature to match the gas temperature closely. The approximation $T(\text{dust}) = T(\text{gas})$ is less accurate at the cloud edges.

Some results are presented in Fig. 12. As expected, the temperature decreases strongly with the density in the inner regions. The low dust temperature (7 K) at the cloud center results in strong modification of the radial emission profile, in particular at 450 μm, which is no longer in the Rayleigh Jeans regime of the blackbody spectrum. Thus, when applied to real cases, such as L1689D or L1544, the derived density profile becomes steeper.

Indeed, when using consistent temperature for L1544, the emission profile may be compatible with a simple $1/r^2$ power law. This is the expected dependency for a singular isothermal sphere, which is also the limit of the (isothermal) Bonnor–Ebert sphere for $\xi \rightarrow \infty$. A potential caveat of this result is the treatment of dust temperature. Two effects are still neglected in most analyses: the dependency of dust emissivity as a function of temperature and the dependency of the temperatures of dust grains as a function of their size.

4.2 Geometry

Another limitation of many existing analyses is the assumed geometry of the source. Spherical symmetry is often used, but real clouds are far from spherical. In some cases, simple modifications can be sufficient. For example, masking the cavity evacuated by outflows prior to radial averaging may be a good first-order approximation.

4.3 Gas Temperature

We expect the gas temperature in a prestellar core to be very low. How does that compare with observations? One can rely on several tools to measure the gas temperature. The easiest would be to observe an optically thick, thermalized transition such as those of the CO isotopologues. Given the expected temperature range, only the two lowest rotational lines can be used. However, this only probes the temperature at $\tau = 1$. For ^{12}CO, this can only probe the outer regions of a core. The rarer isotopologues probe deeper, but can be seriously depleted in the coldest regions because CO sticks to grain surfaces at (dust) temperatures below 17 K. At the expected densities in a prestellar core, this sticking timescale becomes short enough in comparison with the cloud dynamical evolution timescale. Hence, it is difficult to sample the lowest temperature regions with CO isotopologues.

Multitransition studies of other molecules with rotation diagrams or other more sophisticated analysis procedure, as local thermodynamic equilibrium (LTE) does not necessarily applies, can also be used to uncover kinetic temperature behavior. Several difficulties have to be overcome. The first is generic to all source models: unless the lines are at LTE, the relative contribution of density and temperature have to be disentangled. The second is specific to the low temperature conditions reigning in prestellar cores: Only low excitation lines can be used. The third is linked to the depletion onto dust grains, which reduces the amount of molecules and makes them difficult to detect.

Many molecules, such as CH_3CN or CH_3C_2H, which otherwise constitute good thermometers in clouds, are not usable in prestellar cores. One important exception is NH_3, although it is not strictly in the mm domain. The two lowest inversion transitions (1,1) and (2,2) constitute an excellent thermometer. The inversion transitions are easily thermalized and measuring the strength of their emission yields the population in the $J = 1$ and $J = 2$ rotation levels of NH_3, which are separated by 27 K. In case of optically thick emission, the hyperfine structure of the (1,1) transition can be used to measure both the excitation temperature and the opacity of the line, leading to an accurate measurement of the $J = 1$ level population. The same principle also applies to the (2,2) transition, but, in addition, this transition often remains sufficiently optically thin to make the opacity correction negligible.

These properties of the NH_3 molecules have been used by, e.g., [5] to demonstrate the existence of temperature gradients in supposedly stellarless condensations in IC348. Applications to prestellar cores can be found in [28] for L1544. These authors found a rotational temperature increasing from 6 K at cloud center to 8 K at 5000 AU. At these temperatures, this is expected to be very close to the kinetic temperature (see [123]).

A totally independent approach to constrain the temperature is based on the thermal broadening of the lines. As there may be a contribution from systematic or turbulent motions, the total line width provides an upper limit to the kinetic temperature. This is extremely simple, but most molecules are too heavy for this limit to be useful in the temperature range expected for cloud cores. One exception is H_2D^+, whose molecular weight is only 4. This property was used by [65] to constrain the temperature in the ρ Oph D (L1689A) core. A value of 6 K is found.

Multitransition, multimolecule modeling is still the ultimate tool to derive kinetic temperatures. By attempting to reproduce the observed emission from several molecules with transitions sampling different opacity ranges and critical densities, one may be able to disentangle the relative effects of density and temperature. A good example of this kind of approach is given by [97], who studied the L183 (also known as L134N) core. They used high-sensitivity spectra, $12'' - 25''$ resolution of the $J = 1 - 0$ and $J = 3 - 2$ lines of N_2H^+, and the $J = 1 - 0$, $J = 2 - 1$, and $J = 3 - 2$ of lines of N_2D^+. They used a cloud model to reproduce all observed spectra. This requires the use of a 2-D radiative transfer code and appropriate collision rates for these molecules with H_2 (extrapolated from those with He from [30]). Since these molecules have hyperfine structure, the Zeeman sublevels were treated separately to correctly reproduce the relative intensities of the hyperfine components. As these models have many free parameters, such as density structure and molecular abundance, and furthermore are inherently a simplification of the true cloud structure (e.g., the symmetry is never perfect), defining a metric to find some optimal solution is always an issue in such studies. Because of the thermal noise, some observations may have much better sensitivity than others, but these observations may be far less constraining in terms of physical parameters. A purely noise-based normalization can then be totally inappropriate. For example, even a moderately sensitive upper limit on a $J = 3 - 2$ transition can be a better constraint on the temperature than an extremely high signal-to-noise (S/N) ratio spectrum of a $J = 1 - 0$ line. [97] choose to evaluate a reduced χ^2 for some measurements and a direct χ^2 for other measurements, but they present a spectrum-based comparison between the model and observations. They find that the emission is best represented by low temperatures (7 K at the center) and a density structure using two power laws that are steepest at the edges.

4.4 Chemistry: Depletion and Fractionation

Any determination of the kinetic temperature requires some knowledge of the chemical composition of the cloud because at a minimum we need to know where the observed molecules are located. There is ample evidence that the chemistry in prestellar cores is dominated by two major processes: depletion and fractionation.

Images of many molecules have revealed that, in contrast to the dust continuum emission, the emission from molecules is usually not centrally peaked. For example, [118] have imaged the $C^{18}O$ and $C^{17}O$ $J = 1 - 0$, CS $J = 2 - 1$, N_2H^+ $J = 1 - 0$, and $NH_3(1, 1)$ transitions toward several prestellar cores. In all cases, the carbon bearing molecules exhibit a deficit of emission toward the cloud center (see, e.g., Fig. 13). The N-bearing molecules are more centrally peaked, although not as much as the 1.2 mm continuum. This behavior is found in all studied sources so far. A deficit of N_2H^+ emission toward the cloud center has even been observed with interferometric measurements by [11], although their observation was toward a Class I source, IRAM04191, rather than a prestellar core. However, the source is so young that the envelope still behaves as if it were prestellar; in particular, the luminosity is insufficient to heat the dust above 11 K at $R > 400$ AU. In revealing this deficit, it is essential to recognize that short spacings must be added to the interferometric data (see Fig. 14). An apparent "hole" in the interferometer image could be caused by the filtering of extended emission intrinsic to the interferometric data; this hole could even be the result of negative "signal". Thus, all of the molecules appear significantly underabundant in the densest parts of prestellar cores, although N-bearing molecules may appear to deplete less. This may look surprising at first view since N_2 and CO are expected to have similar desorption energy from grains. This specific behavior is, in fact, a consequence of the CO depletion. Consider the reaction

$$N_2H^+ + CO \rightarrow HCO^+ + N_2, \tag{23}$$

which is the main destruction route of N_2H^+. When CO sticks on grains, the equilibrium is modified and relatively more N_2H^+ is left in the gas phase. Ultimately, this enhances the atomic N gas-phase (relative) abundance, leading to NH_3 formation. This appears less depleted than CO in both molecules, although the fraction of N nuclei in the gas phase is actually small compared to those that stick on dust grains.

The second important effect in prestellar cores is deuteration, i.e., the relative abundance enhancement of deuterated species compared to the normal isotopologue. The enhancement is so large that doubly deuterated species, such as D_2CO [18], and even

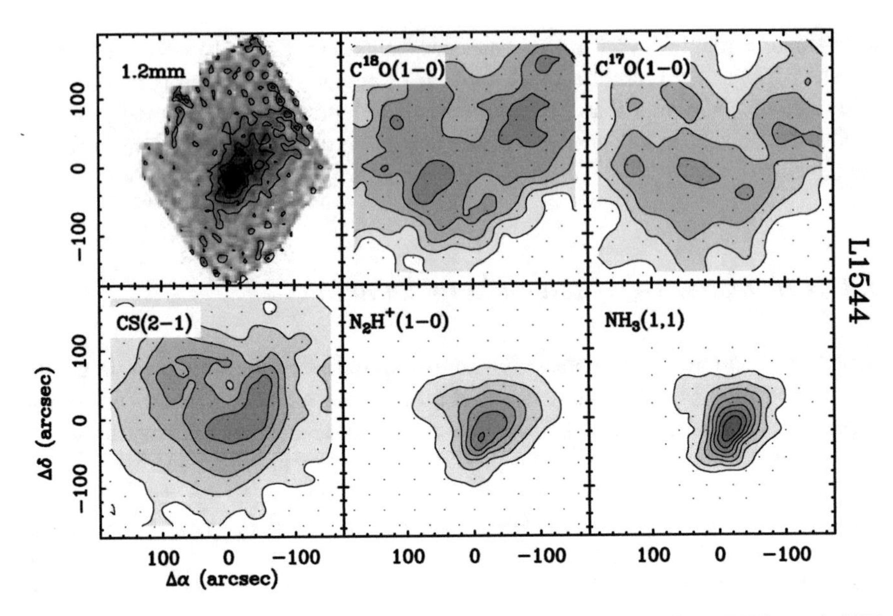

Fig. 13 Images of the emission from several molecules toward L1544 (from Tafalla et al. 2002 [118])

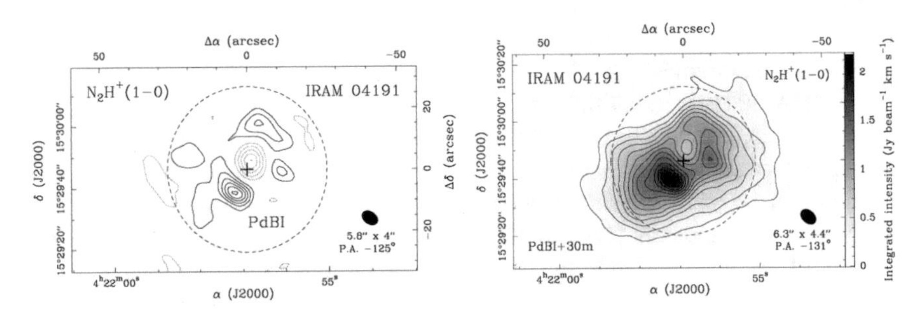

Fig. 14 N_2H^+ $J = 1 - 0$ emission toward IRAM04191 (from Belloche and André 2004 [11])

triply deuterated species, such as ND_3, have been detected in these environments, although the normal abundance ratio of ND_3/NH_3 would only be 10^{-15} (see, e.g., [85, 107]. All these enhancements have their origin in the zero point of the vibrational energy of molecules coupled with the mass difference between the hydrogen and deuterium nuclei. The effect is maximal for H_2 and leads to the following series of exothermic reactions:

$$H_3^+ + HD \rightarrow H_2D^+ + H_2 + \Delta E_1 \tag{24}$$

$$H_2D^+ + HD \rightarrow D_2H^+ + H_2 + \Delta E_2 \tag{25}$$

$$D_2H^+ + HD \rightarrow D_3^+ + H_2 + \Delta E_3, \tag{26}$$

which at low temperature would quickly convert all H_3^+ ions in the deuterated form. However, this chain is normally broken by faster reactions with CO and N_2, i.e.,

$$H_2D^+ + CO \rightarrow DCO^+ + H_2 + \Delta E_c \quad (27)$$

$$H_2D^+ + N_2 \rightarrow N_2D^+ + H_2 + \Delta E_n, \quad (28)$$

unless the abundance of CO and N_2 in the gas phase is very low. The deuteration is then transferred to more complex molecules by other ion-neutral gas-phase reactions. This simplified scheme predicts a correlation between the deuterium enhancement and the depletion of CO in gas phase. This correlation has indeed been observed [8]. Although pure gas-phase processes can account for some of the observed effects [106], deuteration can actually also occur through grain surface chemistry [105], and disentangling the various possible routes involved in this process is one of the challenges of the study of interstellar chemistry. Not all deuteration proceeds through the initial reaction of H_3^+ with HD. For example, the enhancement of DCN over HCN is due to a different chemical path and can occur at higher temperatures.

A generalized version of deuteration is fractionation in which a specific isotopologue is enhanced (or decreased) compared to the elemental isotope abundance. The simplest fractionation effect is the ^{13}CO enhancement occurring at the cloud edges due to the reaction

$$^{13}C^+ + CO \rightarrow C^+ + {}^{13}CO + \Delta E. \quad (29)$$

Although there is no such simple reaction with ^{18}O, other routes exist that can enhance some molecules in specific isotopes (^{18}O, ^{15}N, ^{34}S). In general, the more complex routes are less efficient.

5 Class 0

The study of starless and prestellar cores can provide us with the initial conditions for star formation. The next step is to understand the mechanisms at work once the gravitational collapse has started. This is the so-called Class 0 stage, where the protostar is still surrounded by its parent envelope. In this phase, there are many aspects of theory that should be confronted with reality. Can we see the infall motions? Are they following the pattern expected from the density structure? How is the angular momentum distributed: Do we find evidence for spin up of the core or for counter rotation, which could be indicative of the importance of magnetic fields? When does the protoplanetary disk start building up?

A major difference between this step and the previous stage is the existence of a central heating source. The luminosity rises very quickly and this energy source is expected to have a major impact on the thermal and chemical history of the surroundings. However, as the timescales for collapse are very short (about 10^5 yr), the study of Class 0 objects is difficult because of their (relatively) small numbers.

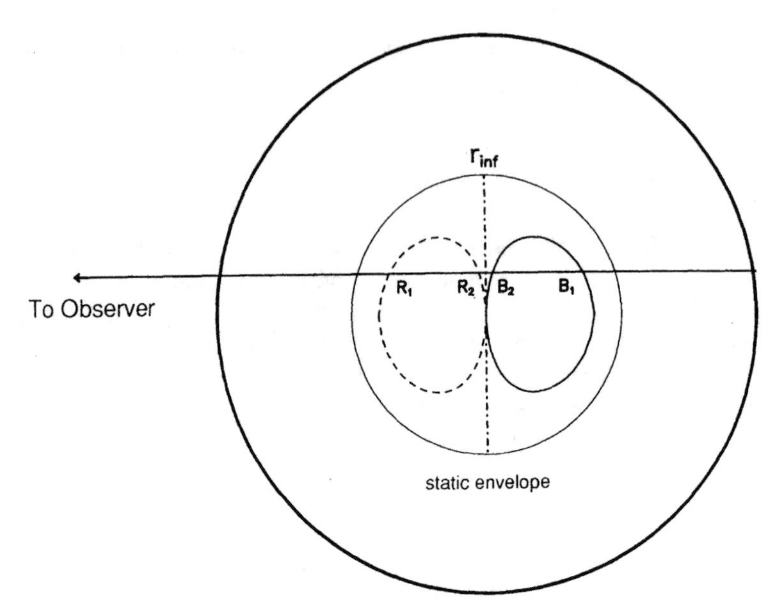

Fig. 15 Schematic model of infalling motions in an envelope (from Zhou et al. 1996 [127]). The solid and dashed lines show contour of equal projected velocity. The dashed line is redshifted and solid line blueshifted for infall

5.1 Infall Motions

The combination of a central heating source and infall motions should result in an asymmetric line shape. Because the radial velocity increases inward, the same projected velocity is achieved in two locations along any line of sight (see Fig. 15).

The temperature and densities are higher at R_2/B_2 than at R_1/B_1 in the case of infall motion. As a consequence, a cold layer hides the warm core for redshifted velocities, while the warm region is encountered first for blueshifted cases. Subthermal excitation can enhance the effect because the density increases in the same manner as the temperature. Accordingly, the overall emission profile displays a redshifted self-absorption dip or, equivalently, an enhanced blueshifted emission hump for optically thick lines. Optically thin lines can be used to determine the systemic velocity accurately. This behavior has been used by [87], who studied the line shapes toward a sample of Class 0 sources. However, the line shapes are more complex than the simple behavior described before. In particular, all Class 0 sources have strong molecular outflows, which can contaminate the line wings. Accordingly, there is no simple measure of the "blueshift" asymmetry. The normalized peak velocity difference $(V_{\rm thick} - V_{\rm thin})/\Delta V$ is ambiguous and sensitive to noise for nearly symmetric lines. An even-odd decomposition is sensitive to the choice of velocity reference. The line centroid difference is sensitive to the velocity window or detection threshold. The ratio of blue-to-red peak brightness is inapplicable if no secondary peak is clearly visible. Line skewness is sensitive to line wings and possibly contaminated by outflows. Nevertheless, [87] chose a normalized peak velocity difference and found an excess of sources showing blueshift asymmetry, as expected if infall motions dominated their sample of $\sim$40 young stellar objects.

Reference [126] observed the B335 core with the IRAM 30 m and showed that the single-dish spectra are well fit by an envelope model, following the inside-out collapse solution of Shu and collaborators [113, 121]. The observed absorption depth depends on three main factors: (i) the angular resolution, as the emission from the envelope fills in the absorption dip; (ii) the line excitation, as higher lying levels are not populated in the envelope; and (iii) the direction, as self-absorption disappears away from the star in a model-dependent way. Factors (i) and (ii) are actually two facets of the same problem.

High angular resolution images, resolving the infall velocity gradient, can thus provide a real test of the theory. An attempt to test the theory for L1527 was performed by [127] with ^{13}CO $J = 1 - 0$ data from BIMA (Berkeley, Illinois, Maryland array) and the NRAO (National Radio Astronomy Observatory) 12 m telescope. Short spacing data are absolutely mandatory in this approach: [55] showed that a *warm emitting layer* can mimic apparent self-absorption when observed with an interferometer because all the large-scale structures are filtered out (see Fig. 16). Adding short spacings is the only way to recover the complete picture. Nevertheless, the combined image of [127] exhibits a patchy aspect, which is the result of difficulties in deconvolving a limited S/N extended structure (whatever the deconvolution technique used); see Fig. 17.

A totally different approach was followed by [95] via the Nobeyama array. This array uses better sensitivity and higher velocity and angular resolution, but no short spacing. The ^{13}CO emission reveals an X-shaped structure, indicating the ^{13}CO traces the edge of the outflow cavity. C^{18}O is elongated perpendicular to the outflow and the velocity gradient reveals a combination of rotation and infall. The overall picture is thus that of a flattened, rotating envelope, infalling toward the star, but with a powerful outflow along the rotation axis that dominates the line wings in the most simple molecules.

A similar study was performed by [55] for L1157 with the IRAM Plateau de Bure interferometer (PdBI), but with the inclusion of short spacings from the 30 m. The self-absorption appears centrally peaked. A comparison of the $J = 1 - 0$ and $J = 2 - 1$ transitions of ^{13}CO constrain the density and, as the CO abundance is rather well known, the overall thickness of the layer since a minimum opacity of 1 is required to produce the self-absorption. This thickness is about 7000 AU, which is several times larger than the apparent width of the layer, and thus indicative of a very flattened, edge-on structure. While this may look a priori unlikely, the outflow from this source is known to lie within 10° of the plane of the sky, however, no rotation is found. These pioneering studies illustrate the main aspect of the structure of Class 0 sources in which a flattened envelope and a collimated outflow are the dominant features.

Another approach to search for infall motions is to attempt to detect absorption *against the continuum emission*. For a solar mass protostar in the earliest stages,

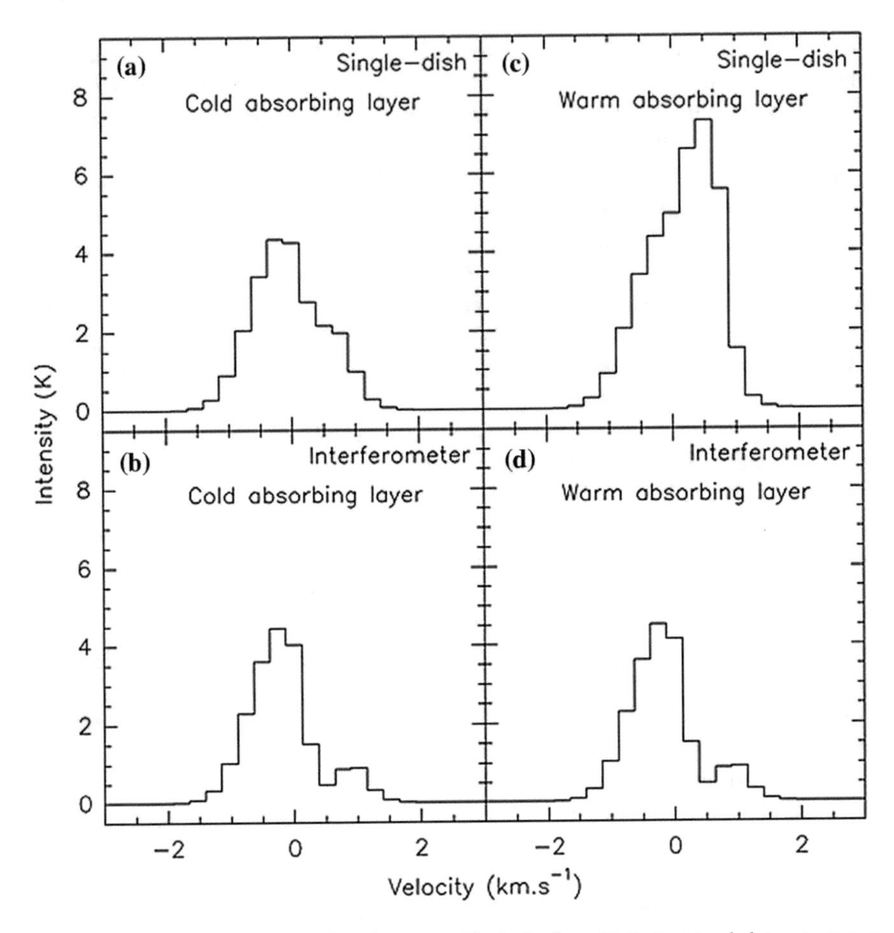

Fig. 16 Filtering effects due to an interferometer. The lack of sensitivity to extended structures can mimic self-absorption even in cases where the actual effect is additional emission (from Gueth et al. 1997 [55])

the mass contained in the inner 50 AU is large enough to produce optically thick continuum emission at mm/sub-mm wavelengths. As this dust is warmer than the surrounding gas, absorption lines can form against this background. High angular resolution is absolutely required to detect such absorption, which can otherwise be hidden by the emission from the surrounding gas. This contamination can also be reduced by selecting a line requiring high densities for thermalization. This has been carried out by [34] for NGC1333 IRAS 4A and 4B. The spectra of H_2CO $3_{12} - 2_{11}$ exhibit P-Cygni profile with redshifted absorption consistent with infall, but can be explained by a simple redshifted absorbing layer. Mapping this absorption is not possible with current instruments, but may become accessible to the Atacama Large mm/sub-mm Array (ALMA).

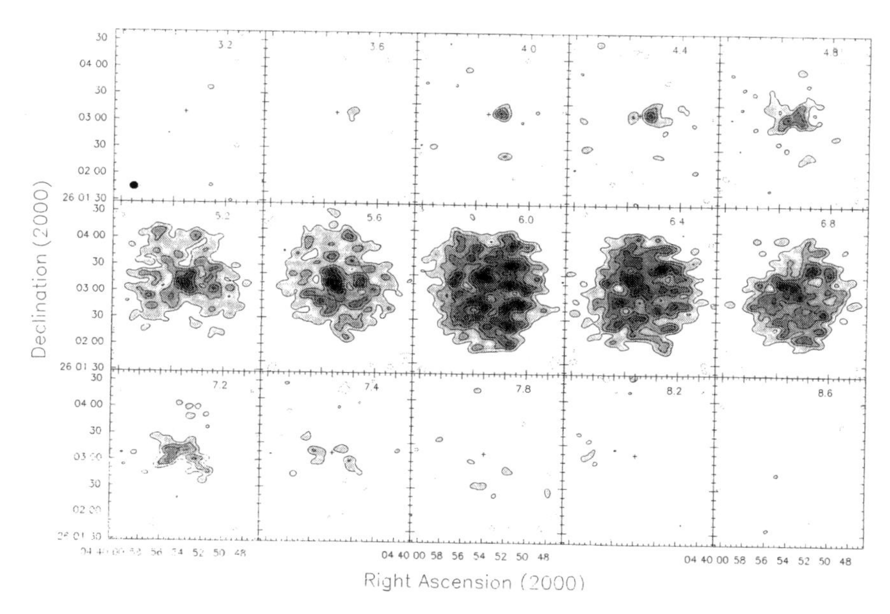

Fig. 17 Typical patchy aspect appearing in deconvolution of limited S/N images even when short spacing is added to recover extended structures (from Zhou et al. 1996 [127])

5.2 Comparing with Models

Detecting infall motions is a first step, but the determination of the overall structure (in density, temperature, and velocity) is required to constrain the mechanism at work. This may allow us to distinguish between alternate theories, e.g., between singular isothermal spheres (SIS) or magnetic models. Images in different tracers, which are sensitive to different regimes of temperature and density, are required to determine these parameters. For B335, [126] attempted to compare H_2CO and CS data with an SIS model by adjusting the infall radius and molecular abundances. The fit is rather poor because of the additional complexity introduced by the existence of the outflow and nonspherical geometry.

The continuum emission from B335 was studied by [67] with the IRAM interferometer. The source is resolved and the visibility curves can be fitted by simple power law. Assuming the source is a spherically symmetric core with a power-law dependence of the density $n(r) \propto r^{-s}$ and $T(r) \propto r^{-q}$, and that the dust emissivity is a (spatially constant) power law of the frequency $\kappa(\nu) \propto \nu^{\beta}$, the emerging intensity in the optically thin, Rayleigh-Jeans approximation is $I(\nu, b) \propto \nu^{2+\beta} b^{-(s+q-1)}$. Neglecting the primary beam attenuation, the Fourier transform of these, i.e., the visibility curve, is $V(u, \nu) \propto \nu^{2+\beta} u^{(s+q-3)}$. The method used by [67] deserves some discussion. In essence, this is a nonlinear least-squares fit using a χ^2 function, but with an additional term to handle calibration errors, i.e.,

$$\chi^2 = \Sigma((V_i - f(u, v; s, m))/\sigma_i)^2 + ((m - m_0)/\sigma_m)^2, \tag{30}$$

where m is the calibration factor for the 1.3 mm data. It is assumed that the 3 mm flux density fixes the overall intensity scale of the problem. The free parameter s is the exponent of the density slope. A common problem in such studies relies on the fact that each individual measurement V_i has an extremely low S/N, but there are many of these measurements (at least several thousands). In such a case, the *reduced* χ^2, $\chi^2_\nu = \chi^2/(N - p)$, where N is the number of measurements and p the number of parameters, is a poor indicator of the fit quality. In fact, as $f(u, v; s, m) \ll \sigma_i, \chi^2_\nu = 1$ even for rather poor models. The value $\chi^2_\nu \neq 1$ can only be obtained for extremely poor models (which are easy to dismiss), or if the noise σ_i has not been properly estimated, which is a much more common situation. One advantage of the large number of measurements is the possibility of using the bootstrap method, in which n random sample of measurements of size N are selected from the original sample and analyzed separately. The distribution of the results provides a fair estimate of the errors, which has the advantage that it is independent of the absolute value of σ. For well-behaved distributions, the result is expected to be similar to analyses of the errors through the covariance matrix. For display purposes, the results are binned (see Fig. 18), but the analyses are best performed on the original data. The observations are indeed well fit by a power law. Overall, [67] find a density index $s = 1.65 \pm 0.15$ that is consistent with the value derived from the SCUBA bolometer data by [112] on larger scales, but that is less than the $s = 2$ index found from extinction studies by [66]. This flatter slope inside would be in qualitative agreement with the behavior of a Bonnor–Ebert sphere or could suggest that a broken power law with $s = 2$ outward could be used (as expected, for example, in the inside-out collapse case). However, it turns out not to be the case because no good fit could be obtained (see Fig. 18).

Although the shape can be matched by the inside-out model, the overall flux seen by the interferometer cannot be reconciled with the larger scale flux. A possible solution to this problem may rely on an evolution of the dust properties with radius, which would change the flux to column density conversion factor.

High-resolution images in line emission are still rare. However, one can use the expected link between infall velocity and distance to the star to provide an effective super resolution with velocity resolved spectra. Given the interplay between chemical composition, density, and temperature, strong constraints can be obtained by observing many transitions from several molecules. High density tracers display stronger line wings. The self-absorption dips depend on opacity, i.e., abundance (therefore, comparing isotopologues is essential) and excitation conditions (as the population of higher lying levels decreases outward). Hyperfine structure is also a useful tool, since the opacity ratios of the hyperfine components can be accurately determined. There are two issues that conspire to make a detailed comparison of the models difficult. First, the problem has too many parameters to be inverted, that is, to recover $n(r)$, $T(r)$, $v(r)$, and the molecular abundances $X_m(r)$ from the observed spectra. The major difficulty here is $X_m(r)$. The appropriate solution seems to be deriving $X_m(r)$ from a chemical model that has known $n(r)$, $T(r)$, $v(r)$. But we are then confronted with the second major difficulty: The chemistry is time dependent. The challenge is

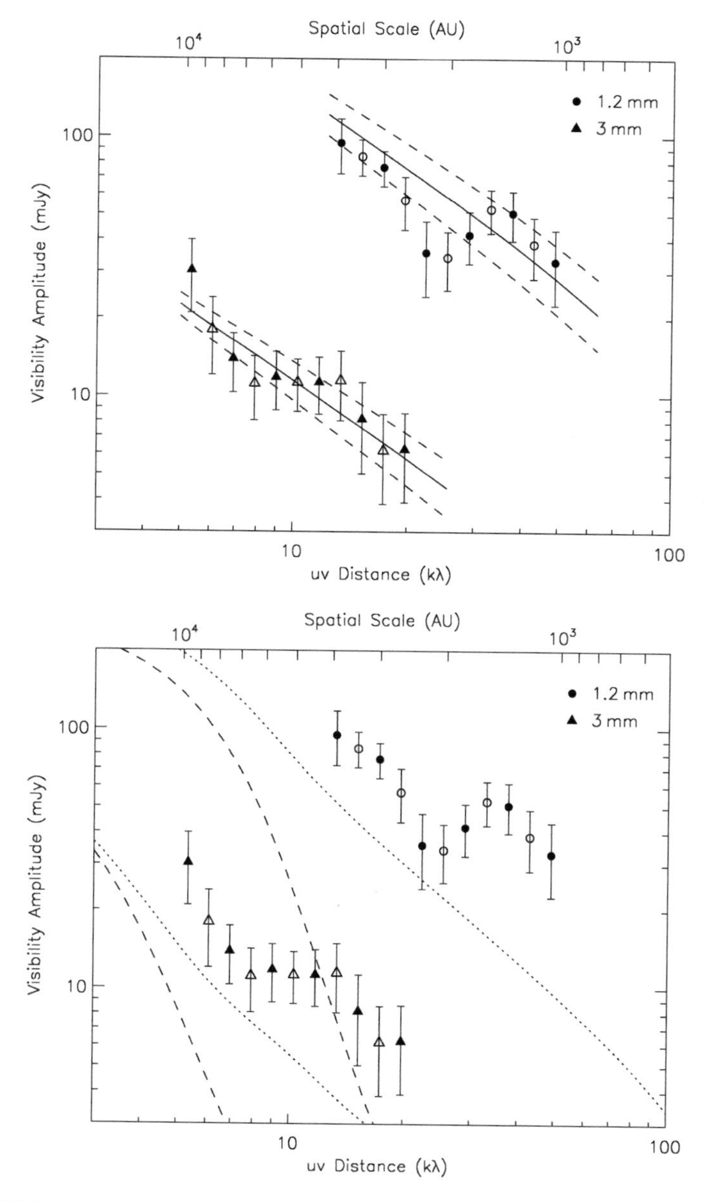

Fig. 18 Visibility curves for the continuum emission in B335 (from Harvey et al. 2003 [67])

then to compute the time dependent chemistry in an evolving (collapsing) structure. [44] attempted to solve these problems in two different ways. In the first solution, these authors prescribed the $n(r)$, $T(r)$, $v(r)$ structure from a theoretical model, which in their case is the inside-out collapse. Then, molecular abundances are simply represented by two regions of constant abundance: the outer envelope and the infalling region. A simple minimization tool can then solve for the best-fit solution to derive these two abundances. [44] used the L_1 norm (sum of absolute differences) rather than the classical least squares (L_2 norm) on the basis that L_1 is less sensitive to outliers, such as the broad line wings, owing to the outflow not considered in the model. This simple model gives good results for some molecules (e.g., CN, N_2H^+), but fails to represent others (HCN). The second approach is a full time dependent solution of the chemistry. The evolution of the density structure is taken as a series of Bonnor–Ebert (BE) spheres with increasing ξ_{max}. The timescale is not well defined at this stage. The BE spheres turn into an inside-out collapse when the SIS stage is reached ($\xi_{max} \rightarrow \infty$), and their evolution follows the model described in [121]. For each model, the dust temperature is computed by solving the radiative transfer and thermal equilibrium. The gas temperature is computed with approximate cooling and heating functions. A time dependent chemical evolution is followed throughout the process. This implies that initial conditions may play a role. [44] use, as a starting condition, a cloud that was initially in atomic form (H_2, He, He^+, C^+, N, O, and metal ions), and whose chemistry has been evolving for 1 Myr before the sequence of BE spheres is activated. There is no dust surface chemistry in the whole process, but molecules are allowed to stick on grains. Models were run for a few varying parameters: the dust grain surface type (which controls the binding energy), the sulfur abundance, and the cosmic ray ionization rate. Despite these approximations, the overall agreement in the best-fitting set of parameters is much better than with the simple step function for the abundance.

The most important disagreement with the infall models remains the velocity of the self-absorption dip. In the inside-out collapse, the outer envelope is at rest and, thus, the strongest self-absorption dip is at the systemic velocity. In contrast, observations find a redshifted absorption dip, suggesting the presence of an extensive layer infalling toward the source.

6 Outflows

Outflows are ubiquitous in protostars and progressively disappear in the Class II stage. In particular, outflows represent a major issue in our understanding of the star and (perhaps) planetary formation process. We would like to understand what fraction of the angular momentum is actually carried away by outflows, what is the corresponding mass loss history, and how they may participate in the parent cloud disruption. Understanding the thermal and chemical history and the dust evolution in outflows is a challenge. Identifying the origin of outflows is of critical importance: Current results suggest that outflows originate from within a few AU of the parent star and may thus severely affect the planetary formation process in such regions.

Outflows also represent a nuisance in understanding the properties of the surrounding medium. The mere existence of outflows proves that there is no spherical symmetry. Yet, most analyses presented previously were relying on 1-D spherically symmetric models. Also, we must avoid confusing outflow signatures with other fundamental features of the star formation process: infall motions, high-velocity Keplerian wings, and "hot cores".

Outflows are revealed by high-velocity wings of spectral lines at velocities large enough to make it unlikely (or even impossible) to trace gravitationally bound motions. Outflows are most easily visible using CO, and are in general bipolar and collimated, but exhibit wide variations in collimation angles. Figure 19 shows typical outflows from low-mass stars seen by the Owens Valley Radio Observatory (OVRO) interferometer [3]. The opening angle appears to increase with time since outflows from Class 0 sources are extremely collimated. The opening angle is also a function of velocity. Images of high-velocity wings, in general, display a jet-like morphology, while low-velocity components delineate the edges of a cavity with a conical opening near the exciting source, which connects to bow-like structures at the flow apex. A typical example is HH 211 [53]; see Fig. 23.

Although highly collimated, outflows often extend over considerable distances, up to a few parsec from the exciting sources. These overall properties make outflows rather difficult to study.

The high collimation often implies that only interferometers can resolve them transversally, but the large span on the sky requires the use of single dishes and, with interferometers, mosaicking techniques (at least in one dimension). In the latter case, it is often necessary to add in short spacings from single-dish observations. This is especially important for flows near the plane of the sky. The lack of short spacing information can be critical for low-velocity gas, which in general represents the bulk of the mass in the outflow. The need for short spacing information is well illustrated by the study of the L1157 outflow [54]; see Fig. 20.

6.1 L1157

The main properties of outflows are indeed well illustrated by the example of L1157. The southern lobe has been studied with mosaicking techniques using the IRAM Plateau de Bure interferometer [54]. In this source, the high-velocity CO gas traces two expanding cavities, perhaps caused by two separate main ejection events. A simple kinematic model including two conical cavities allows us to reproduce most of the apparent features (see Fig. 20). In their models, [54] found that the velocities were nearly radial. However, the velocity perpendicular to the flow axis is found to be larger than the velocity along the flow axis times the cavity aspect ratio. This implies that the structure broadens with time, i.e., a time decreasing collimation. The inclination is well constrained by the model with the flow nearly in the plane of the sky $i = 81°$. The misalignment of the two cavities and their different (projected) velocities suggests that the overall ejection process responsible for the cavity creation is accompanied by a precession of order 6° with a period about 4000–5000 years.

The cavities appear nearly empty, however, this result should be treated with some care because the observations are based on the CO $J = 1 - 0$ line alone. The cavity

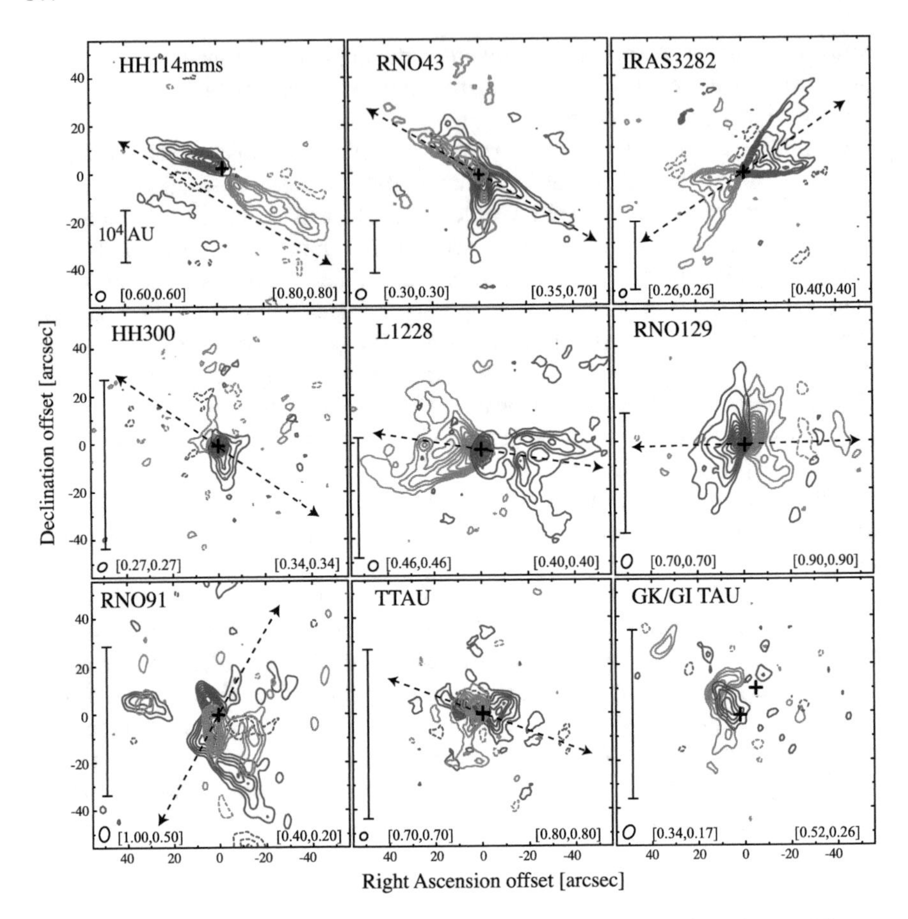

Fig. 19 A gallery of outflows seen in CO with the OVRO interferometer (from Arce and Sargent 2006 [3])

could well be partially filled by warmer, higher velocity gas to which the $J = 1 - 0$ observations are not sufficiently sensitive. Observations of high J transitions of CO are unfortunately often missing for outflows.

A curious property of the L1157 outflow is a marked asymmetry between the two lobes. The northern lobe has been imaged with the 30 m telescope; see, e.g., Fig. 1 of [7]. This lobe displays an overall morphology that is point symmetric to that of the southern lobe; however, the northern lobe is systematically larger by about 50%. This scale difference also appears in the outflow velocities. All this suggests that the morphology of the outflow is driven by a precessing jet moving about 50% faster toward the north. The origin of this velocity difference must lie very close to the exciting source, as the scale factor is rather homogeneous.

An underlying high-velocity jet is indeed revealed by other molecular tracers, of which the most spectacular is SiO. The abundance of SiO is enhanced by orders

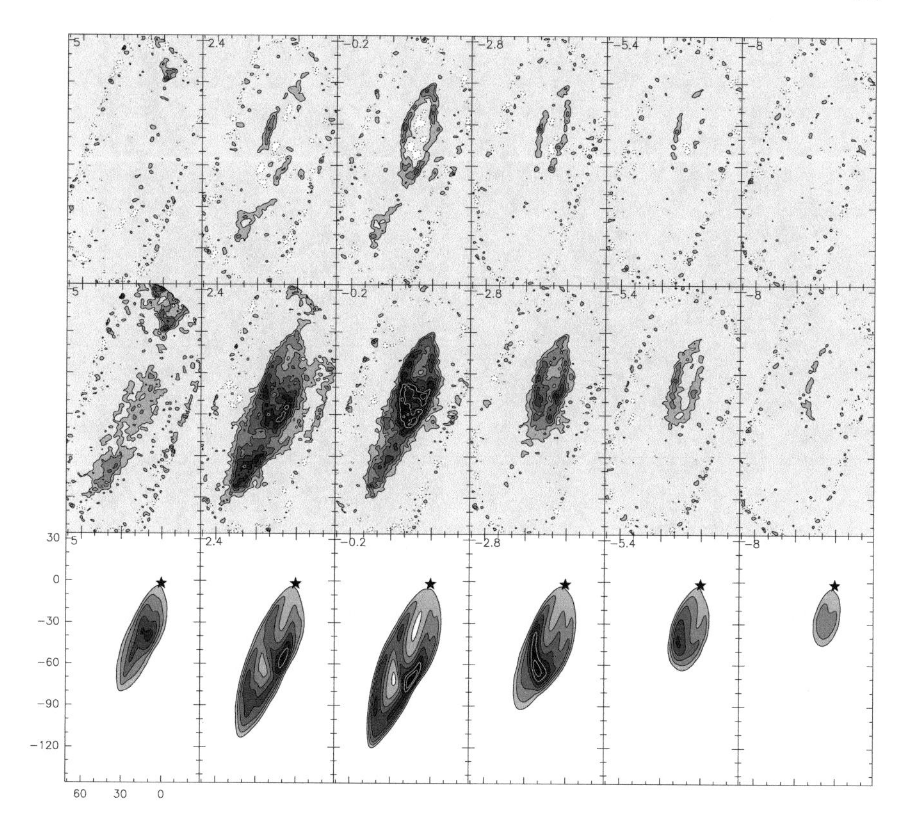

Fig. 20 An illustration of the importance of the short spacings for the study of outflows. Top: CO image of the L1157 outflow from the IRAM interferometer data only. Middle: The same outflow after adding short spacing information from the 30 m telescope. The "bar" in the middle of the top cavity is totally invisible in the interferometric data alone (from Gueth et al. 1996 [54])

of magnitude at the bow-shaped apex of the main cavity and in a jet-like feature emerging from this cavity (see Fig. 21). This enhancement of SiO abundance is most likely due to sputtering of Si from grains in the shocks, followed by chemical reactions leading to the stable form of SiO [109]. SiO appears at higher velocities than CO, and simple bow-shock models suggest that it could be the result of 20 km s^{-1} shocks in a medium traveling at about 50 (main cavity) or 25 km s^{-1} (outer cavity). The enhancement of SiO abundance at the western edge of the inner cavity is consistent with the precession model suggested by CO only because the model predicts that this edge is currently impinged by the high-velocity precessing jet.

Although it is the most spectacular, SiO is not the only molecule affected by shocks. Indeed, the chemistry of the L1157 flow is extremely rich, and many molecules have been detected in this outflow. The initial detections were made using the 30 m telescope, taking advantage of simultaneous observations of many molecules and transitions at the same time [4]. Figure 22 represents spectra of various molecules

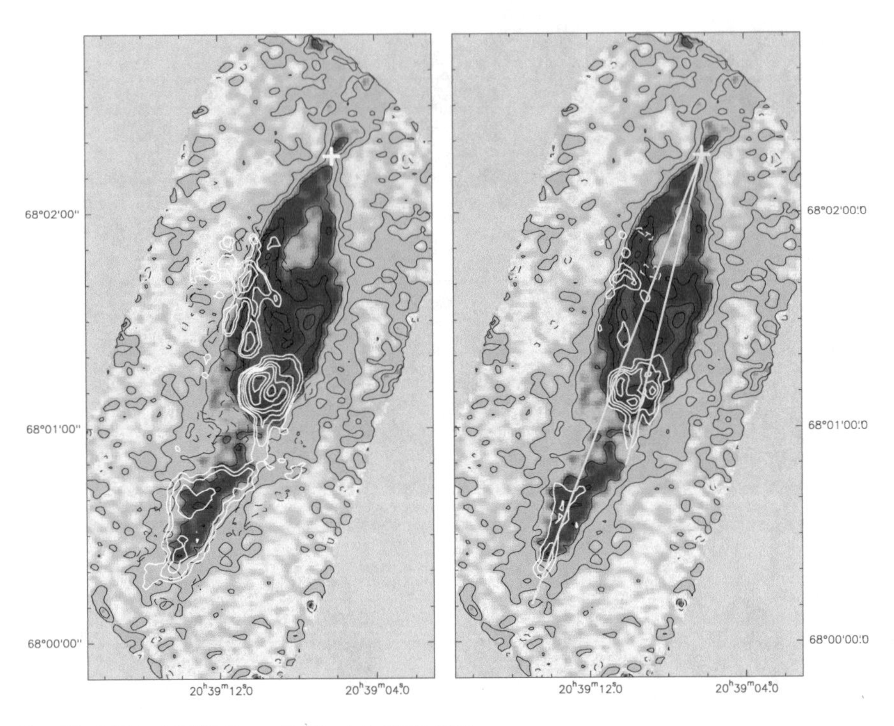

Fig. 21 SiO in L1157 (from Gueth et al. 1998 [56])

toward the protostar and the apex of the two main southern cavities. The following molecules are only detected in the outflow: CO, SiO, CH_3OH, H_2CO, CN, HC_3N, CN, SO, SO_2, and OCS. Others, such as N_2H^+ and C_3H_2, only display narrow lines, which are only detectable toward the protostar, as is also the case for DCO^+. A few molecules have a hybrid behavior: HCO^+, for example, is clearly visible with both the wide and narrow velocity components. Ammonia also mostly traces the protostellar envelope in the (1,1) inversion transition, but only the outflow in the (3,3) line [117].

Mapping with the 30 m reveals clear differences in the spatial distribution of these species. Observing several transitions of the same species enables us to disentangle abundance variations and excitation effects due to temperature and density changes [7].

Multiline analysis can be done via, for example, a rotation diagram method or a classical large velocity gradient approximation. However, because of the complex morphology, single-dish observations do not resolve the expected strong gradients (in density, abundance, and temperature). Accordingly, the derivation of mean values from such unresolved observations can be severely misleading and should be considered as a qualitative indication. From JCMT (James Clark Maxwell telescope) observations of the CO lines up to $J = 6 - 5$, [68] derived temperatures ranging up

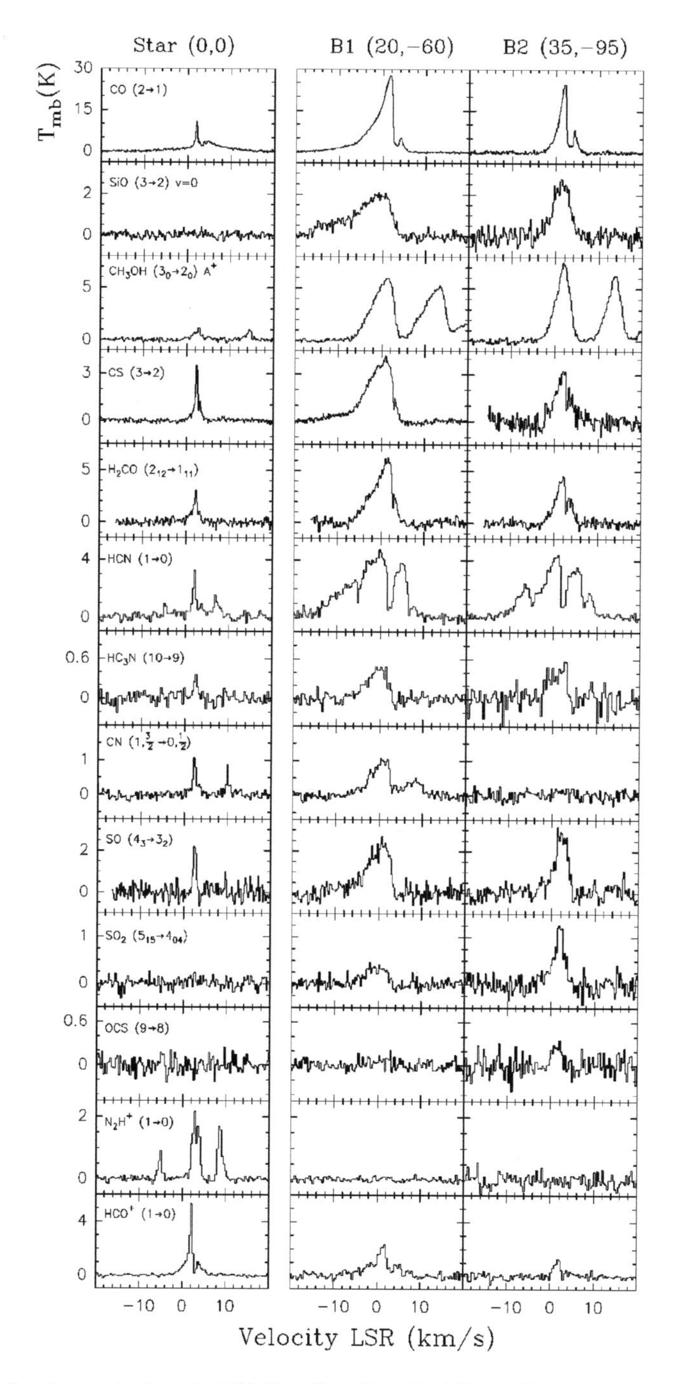

Fig. 22 Molecular spectra from the L1157 outflow (from Bachiller and Perez-Guttierez 1997 [4])

to 180 K with a clear dependence on the temperature as a function of gas velocity (the higher, the warmer).

Interferometric observations have started to reveal the details of the morphology of the emission of a few molecules [13, 55], showing strong variations between, e.g., CH_3OH, CS, and SiO. The interpretation of these differences rely on the different chemical behavior of these molecules in shocks and as density tracers. CH_3OH can be simply removed from grain mantles by relatively slow shocks, while SiO requires more efficient sputtering of Si from grain cores, so the relative abundances of these molecules depend on shock velocity. The post-shock density also plays a role in the line excitation. The precessing configuration of the L1157 facilitates strong variations of incidence angle, and hence transverse shock velocity with position, which results in differences in post-shock density and temperatures that strongly affect the chemistry. Even with the detailed images provided by the interferometer, these variations remain partially unresolved.

Most interferometric observations of L1157 do not include short spacings (e.g., in [13], only HCN and CH_3OH have short spacings added). The importance of this omission can be partly quantified by comparing the integrated spectra derived from the interferometric data (after deconvolution) to single-dish spectra obtained, for example, by the 30 m telescope over the same region. The missing flux (expressed in brightness units) should then be compared to the rms noise of the interferometric map. If the missing brightness is at most of order the rms noise, adding the short spacing information is not critical.

The last important feature revealed by the observations of L1157 concerns the dust properties. Using bolometers (at 1.2 mm with the IRAM 30 m telescope and 0.85 mm with the JCMT), [58] detected continuum emission from the outflow. Although the strong lines owing to the enhanced molecular abundances and temperatures in the outflow contaminate the bandpass of the bolometers, the bulk of this emission is attributable to thermal dust emission. The intensity ratio between 0.85 and 1.2 mm indicates an emission spectral index $\alpha \approx 4$ ($\alpha = 2 + \beta + \gamma(T_d)$ in the optically thin regime, where $\gamma(T_d)$ is a correction for the Planck function at the dust temperature T_d, $\gamma(13\,\mathrm{K} = -0.6)$ in the direction of the main cavity; this is indicative of small dust grains. In contrast, $\alpha \approx 2 - 3$ toward the protostar. This low value indicates some grain growth compared to typical interstellar dust grains, however, it is possibly affected by some opacity effects because the flattened circumstellar envelope is seen nearly edge-on, which is a situation that maximizes the opacities.

6.2 Other Representative Cases

Although L1157 displays most (if not all) of the characteristics of outflows from young (proto)stars, this star is atypical in terms of its chemical richness. In fact, L1157 may represent a peculiar stage of the outflow evolution in which the chemical reactions driven by the shocks only deliver the maximum abundances for many molecules. Younger outflows are more collimated and, in general, faster; on the other hand, more evolved outflows widen and slow down; and both cases lead to a different (and less rich) shock chemistry. Good examples of younger outflows are those driven by L1448-mm and HH211-mm.

L1448-mm

In L1448-mm, interferometric observations of SiO [61] revealed a high-velocity jet entraining a CO outflow with a typical cavity morphology that is very elongated [6]. A complete mosaic of the southern lobe in SiO showed that this molecule may trace large bow shocks in the CO cavity [40]. Proper motion of the bright spots in SiO have been measured by comparing images taken at two different epochs with two different arrays [47]. The derived proper motions, about $0.1''$/yr, combined with the projected velocity, enable a direct measurement of the outflow inclination; the derived value, $\sim 21^\circ$, is in excellent agreement with the inclination determined by adjusting the projected shape of the cavity as a function of projected velocity using a simple geometric model. This also allows us to measure the true jet velocity, which is around 180 km s^{-1} in this case. In this outflow, the extremely high-velocity gas is found to be very warm with temperatures ranging up to 500 K.

The magnitude of the expected proper motion, $0.1''$/yr for typical jet velocities (150 km s^{-1}) and source distances (200 pc), is large enough to be non-negligible when merging data taken at different epochs. This is an additional complexity for high angular resolution studies with current instruments, as obtaining all the required configurations may require a time span of more than a year. The situation will improve with ALMA, but we will still face the issue that there may be significant morphology changes with time. L1148 also illustrates an important feature of outflows: the effect of the surrounding environment. The northern lobe appears bent compared to the average flow axis. Reference [6] attribute that to a physical interaction (collision) with another outflow from the nearby source L1448 IRS3.

HH211-mm

HH211-mm is perhaps an even younger outflow. It was discovered through near-IR imaging in the H_2 2.12 μm line by [89]. Figure 23 shows an overlay of the CO $J = 2 - 1$ emission and low and high velocities, superimposed on a gray scale image of the H_2 emission. The flow was imaged in CO $J = 1 - 0$ and $J = 2 - 1$ by [53] in mosaicking mode with the IRAM interferometer. As the flow is extremely collimated, the mosaic is linear. No short spacings were added, but the missing flux has been estimated with 30 m single-dish spectra, and corresponds in the most affected low-velocity channel to about 1 or 2 contour levels in Fig. 23. The shape of the main cavity is well represented by a simple analytical model of bow shock propagation as developed by [103, 124]. This is however a simplification. Small differences appear near the exciting source and a second cavity is found with a different axis beyond the main bow shock, suggesting a sudden change in flow direction at this stage. The kinematic age of the flow is very small, less than a 1000 yr for the main bow shock. The high-velocity gas delineates a very highly collimated jet. The blue and redshifted sections are misaligned by a few degrees (3°). This misalignment may be intrinsic to the ejection mechanism, but it could also be caused by transverse

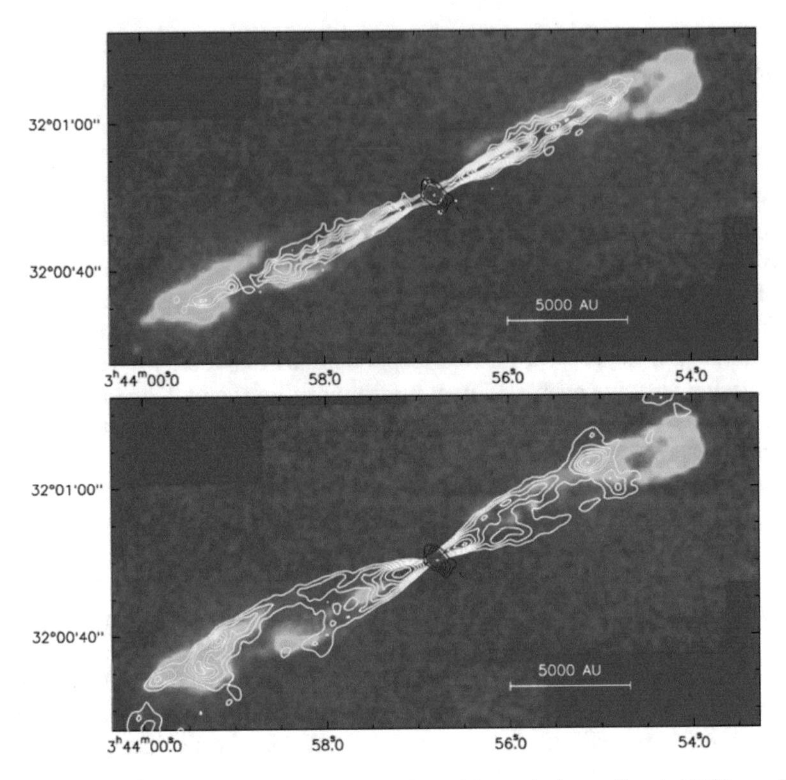

Fig. 23 Jet-like and cavity morphology of the high- and low-velocity outflow in H211 mm (from Gueth and Guilloteau 1999 [53])

motion of the exciting source in the plane of the sky at about 2 km s^{-1}, if the flow material is losing this transverse velocity because of pressure forces from the ambient material. Evidence for such deflection resulting from pressure effects has been found in neighboring flows from large-scale, near-IR images [122].

The HH211 jet is beautifully seen in SiO. Highly excited lines have been detected (up to $J = 8 - 7$, see [98]). High angular resolution (at the arcsecond level) of the SiO $J = 1 - 0$ [20], $J = 5 - 4$ [69], and $J = 8 - 7$ allow us to constrain the excitation conditions in the jet. Densities (of order 10^7 cm^{-3}) and temperatures (up to 300–500 K) are very high near the source and decrease with distances. The SiO emission is not continuous, but is concentrated in prominent knots that are presumably internal shocks in the jet. The SiO emission peaks at velocities larger than those of CO, suggesting the emission is more closely related to the primary jet. The interpretation of the kinematics remains under debate. At limited signal to noise and resolution, a linear correlation between velocity and distance from the exciting source (the so-called Hubble flow law) dominates the position-velocity diagram [69]. Such behavior is found in jet driven winds (e.g., [84, 114], but not in entrainment mechanisms (of course, a Hubble law can simply result from ballistic motions from a single event with a velocity dispersion). However, higher sensitivity and angular resolution observations of the $J = 5 - 4$ line with the IRAM interferometer reveal a different picture. While the strongest emission in SiO indeed peaks at velocities proportional to the distance to the star, there is also significant emission at larger velocities, up to 30 km s^{-1} in projection (the inclination angle is unknown, but likely ranges between 10 and 30°). The very large velocity dispersion at any point is more suggestive of the transverse motions resulting from internal bow shocks propagating in the jet.

Although L1157, L1448-mm, and HH211 illustrate some of the properties of outflows, there is a huge diversity in these properties and the main driving mechanism is still a debated issue. A particularly puzzling issue is why some outflows appear single-sided, even though they surround bipolar optical jets as in the case of HH30 [99]. The nature of shocks in outflows is also unclear, in particular, the importance of magnetic fields in leading to C ("continuous") or J ("jump") shocks. The C shocks should be affected by a magnetic precursor, and one of the tools to reveal the existence of these shocks is the ion drift with respect to neutrals, which is expected in this precursor. A pioneering study has been made by [75], revealing two velocity components in molecules that may probe the existence of a magnetic precursor (HCO^+ and SiO). Further studies by [76] demonstrated a strong enhancement in ionization degree in the precursor region. The increased electron density results in much stronger excitation for molecular ions than for neutral molecules in this region because ions have larger collisional cross sections with electrons than neutrals.

All these results have been obtained despite serious limitations: limited angular resolution, no or few mosaics, practically no quantitative analysis, and unclear geometry. Many of these limitations will be removed by ALMA, in particular, owing to its ability to produce high-quality mosaics.

7 Protoplanetary Disks

As a result of the conservation of angular momentum, the formation of a rotating disk, which ultimately reaches a Keplerian pattern due to viscous forces in the gas, is a natural outcome of the protostellar collapse. These so-called protoplanetary disks represent a key stage in the formation process of planetary systems. Thirty years ago, "Do disks exist?" was still a valid question. Although the IR SED suggested the existence of flattened structures around young stars, millimeter telescopes and, in particular, interferometers have played a key role in resolving this question, and even in answering the subsidiary question, "What is their size?". With the increasing number of known exo-planetary systems, the main questions we are now asking concerning such disks are:

- How much gas and dust do they contain? Is there enough to form a planetary system?
- How do dust grains evolve? Can we detect evidence of grain growth and sedimentation?

- Where and when do planets form? Can we find evidence of forming planets?
- When and how does the gas disappear? Is the gas disappearance the result of accretion, photo-evaporation, planet formation, or a combination of these?

To answer these questions, we must find the tracers of disk evolution.

In the earliest phases, mm/sub-mm arrays provide the best tools to study circumstellar disks. Early studies revealed that disks are much larger than initially believed from the models of our solar system formation (up to 800 AU radius for GM Aur, DM Tau, and GG Tau, [38, 59, 78]). Their masses are significant (a few % of the stellar mass), and as a consequence, the dust optical depth through the disk is large. Only long wavelength observations can penetrate through the disk. The mm/sub-mm domain is well suited, as the dust becomes optically thin (at least beyond 30 AU or so) at wavelengths >1 mm. Furthermore, the disks are cold, <20 K beyond 100 AU, so molecules and dust only emit in the mm domain. Optical telescopes and IR studies (especially IR spectroscopy) still provide useful complementary information, however, because of the dust optical depth they only sample the upper layers. These wavelength domains become progressively more interesting when the disk dissipates, as the optical depth decreases.

The key ingredients in protoplanetary disks are best illustrated by HH 30. Figure 24 shows a flared, edge-on disk revealed in scattered light by the HST [17]. The thermal emission from the disk has been imaged with the IRAM interferometer [62]. The ^{13}CO $J = 2-1$ line reveals the disk rotation, while the more optically thick ^{12}CO $J = 2-1$ transition traces an outflow cone [99], surrounding the optical jet detected by the HST.

Disks are heated by the central object through their surface and their gravity is dominated by the central star. The dynamic structure is in a quasi-steady state; the angular momentum carried inward by the (low) accretion rate is compensated by a slow outward spreading. The flaring is a result of hydrostatic equilibrium, where the combination of the (slowly decreasing) temperature and Keplerian rotation results in a hydrostatic scale height increasing faster than r with radius.

7.1 Simplified Disk Parameters

To understand how mm observations can constrain the disk physics, it is useful to implement the simplifying assumption of power-law distributions for the main disk parameters. These parameters are (i) the surface density $\Sigma(r) = \Sigma_0(r/r_0)^{-p}$; (ii) the rotation velocity $V(r) = V_0(r/r_0)^{-v}$; and (iii) the temperature $T(r) = T_0(r/r_0)^{-q}$. In the absence of vertical temperature gradients, the vertical density structure is a Gaussian and its scale height iv) $H(r) \propto T(r)^{1/2} r/V(r)$ is also a power law, $H(r) = H_0(r/r_0)^{-h}$ with $h = -(1 + v - q/2)$. We stress that two conventions coexist to define it. Theorists tend to specify $H(r) = C_s/\Omega_{\mathrm{kep}}$, while observers define the Gaussian as $n(r, z) = \Sigma(r)/(H(r)\sqrt{\pi}) \exp(-(z/H(r))^2)$, i.e., $H(r)$ is the $1/e$ half width of the density distribution, which is $\sqrt{2}$ larger than the previous convention.

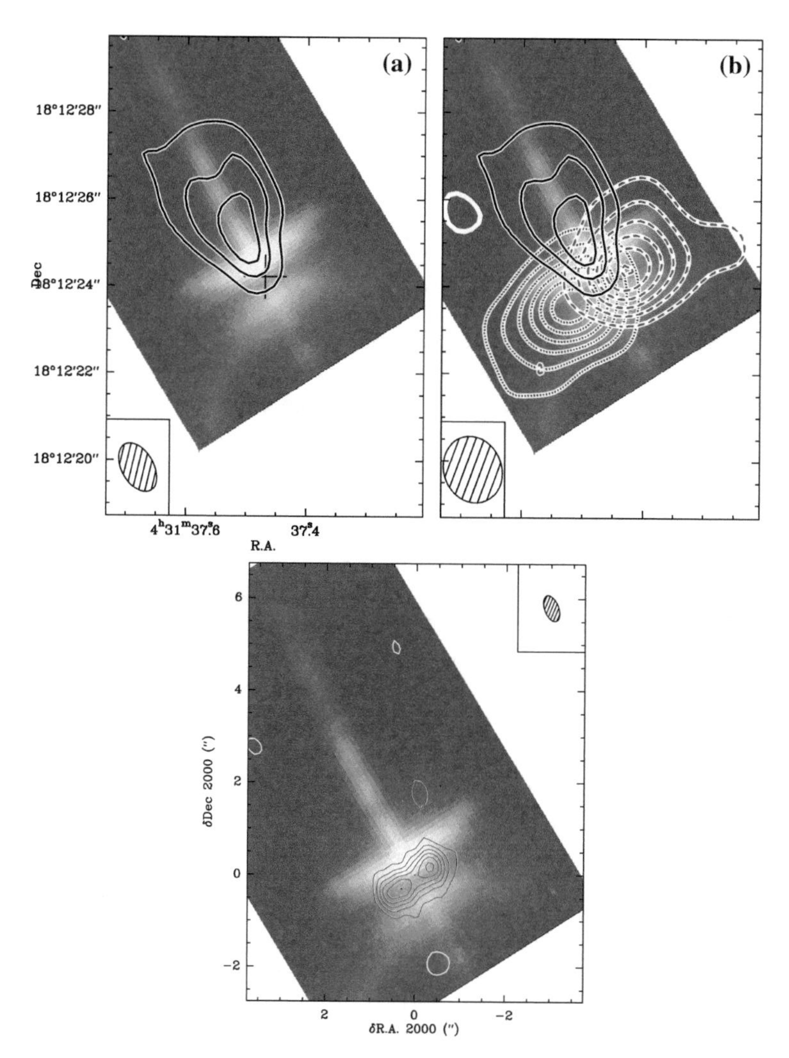

Fig. 24 The archetypical object HH30, a highly embedded young star hidden by an edge-on dust disk (seen in scattered light in false color). An optical jet emerges from the central object and is surrounded by a molecular flow. The Keplerian rotation of the disk is traced by 13CO emission. High-resolution observations of the dust emission at mm wavelength reveal an inner cavity, which is most likely due to the binary nature of the star (from Pety et al. 2006 [99] and Guilloteau et al. 2008 [62])

The power-law approximations allow us to derive constraints independently from any detailed (but always uncomplete) physical model because they are the linear approximations in a log/log domain. Our a priori knowledge indicates that these power laws indeed represent a good first-order approximation. A power law is indeed an exact representation of velocity as a function of radius with an exponent $v = 0.5$ for Keplerian rotation. It is also a good representation of the dust and gas temperature

[25]. For surface density, the paradigm of quasi-steady state disks with a constant α viscosity [111] predicts

$$\alpha C_s H(r) \Sigma(r) = \dot{M}/(3\pi), \tag{31}$$

which, given the definition of the relation between sound speed and scale height, leads to

$$\Sigma(r) = \frac{\dot{M}}{3\pi\alpha} \frac{r}{H(r)^2 V(r)}. \tag{32}$$

This does not hold for the disk edges, however. Self-similar solutions taking viscous spreading into account (under the assumption that the viscosity is power law of the radius and constant in time) predict an exponential edge on top of the following power law [86]:

$$\Sigma(r) = \Sigma_0 \left(\frac{r}{R_0}\right)^{-\gamma} \exp\left(-(r/R_c)^{2-\gamma}\right). \tag{33}$$

From Eq. 33, it can be derived that R_c is the radius containing about 63% of the disk mass. Depending on whether R_c is small (which could be expected for young disks) or large, power laws may also be good first-order solution.

For dust emission at mm wavelength, power laws may thus be a good approximation, provided the dust emissivity varies smoothly with radius. For a given molecule, the molecular abundance and line excitation have to be considered for the emissivity of each transition. Power laws may in these cases be poor approximations.

An interesting limiting case is that of the $J = 1 - 0$ rotational transition of simple molecules. As the disk densities are high, in general, these are thermalized. The densities in disks are so high that the lowest rotational levels are likely to be all thermalized, leading to an effective "rotation" temperature equal to the kinetic temperature. A simple application of the equations given above (Sect. 3) indicates two useful limiting cases for the fundamental rotational transitions of molecules. In the optically thick case, the line surface brightness is just the kinetic temperature of the emitting region. In the optically thin case

$$T_b = (1 - \exp(-\tau))T_k \quad \approx \tau T_k \quad \propto \Sigma_m(r)/T_k(r)/\delta V(r) \tag{34}$$

because the opacity goes as $1/T_k^2$ at high enough temperatures. The brightness should thus fall as a power law with exponent $p - q$ for constant local line width.

7.2 *Disk Parameters from Spectroscopic Observations*

The Basics of Spectroscopy in Disks: Super-Resolution

An essential property of protoplanetary disks is the power-law relation between rotation velocity and radius. Figure 25 shows the area of equal line-of-sight velocity

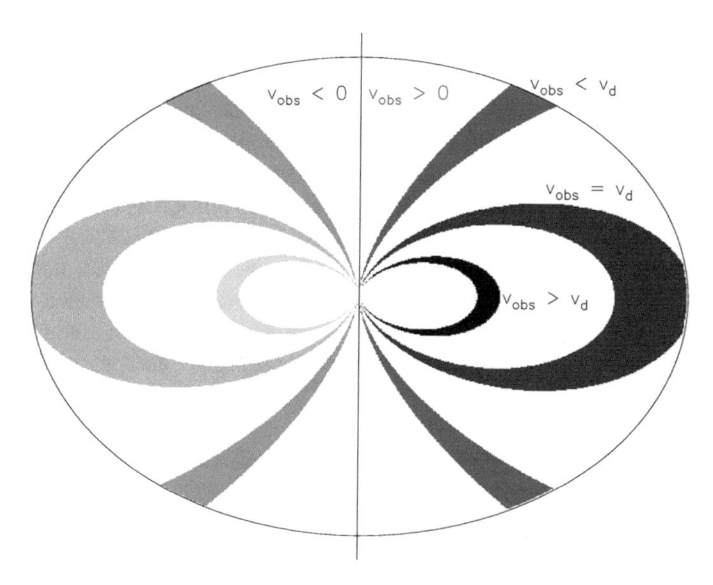

Fig. 25 Loci of equal line-of-sight velocity for an inclined Keplerian disk

for an inclined Keplerian disk. The locus of points of projected velocity v is given by

$$r(\phi) = GM_*/v^2 \sin^2(i) \cos^2(\phi). \tag{35}$$

The largest emitting area is obtained for $V_d = V(R_{\rm out} \sin i = \sqrt{GM_*/R_{\rm out}} \sin i$ and it covers a fraction of order $\delta V/2V_d$ of the disk, where δV is the local line width. Accordingly, the peak flux cannot exceed

$$S_\nu = S_{\rm thick} \frac{\delta V}{2V_d} \cos i, \tag{36}$$

where $S_{\rm thick}$ is the flux of the optically thick disk of same size, e.g., for a constant temperature

$$S_\nu = \pi \left(\frac{R_{\rm out}}{D}\right)^2 \frac{2kT}{\lambda^2} \cos i \frac{\delta V}{2V_d}, \tag{37}$$

and, as the line width is $\sim 2V_d$, the integrated line flux of an optically thick line is of order

$$\int S_\nu dv = S_{\rm thick} \delta V, \tag{38}$$

i.e., proportional to the *local* line width. For an optically thin line,

$$\int S_\nu dv \approx \tau S_{\rm thick} \delta V, \tag{39}$$

the opacity τ goes as $\Sigma/\delta V$, so the line flux only scales with the total number of molecules as expected.

A second important property of protoplanetary disks is that the surface density, in general, drops much faster with radius than temperature; more precisely, in the power-law approximation, $p - 2q > 0$. In these circumstances, even the low-lying rotational transitions of simple molecules have their maximum opacity at small radii (as the opacity scales as Σ/T^2 because of the partition function). Thus, the emission at high velocities, which comes from the inner region of the disks, may be optically thick. Under these circumstances, considering the relation between radius and velocity, it can be shown that the intensity in the high-velocity line wings scales as $(v/v_d)^{(3q-5)}$ [10, 70]. If the line is optically thick up to the outer radius, it can also be shown that the low-velocity portion of the line profile scales as v/v_d, provided $v > \delta V$.

These simple geometric properties enable an accurate retrieval of many of the disk parameters even with limited angular resolution because the spectral resolution provides a proxy for angular resolution.

Understanding a Velocity-Resolved Image of a Disk

Considering the above properties, it is easy to understand how the disk parameters can be recovered from images as a function of velocity for a given spectral line. The shape of the emission as a function of velocity accurately constrains the inclination and exponent of the velocity law v, which must be 0.50 for Keplerian rotation. With arcsecond resolutions, the inclination can be constrained to within 1° and the exponent v to about $0.01 - 0.02$. The width of the emission (the spatial extent perpendicular to the location of equal projected velocity) is proportional to the local line width δV. In the optically thick central regions, the radial dependence of the brightness gives the temperature dependence and constrains q. In the outer regions, the intensity depends on both the (molecule) surface density and temperature. However, as the temperature is constrained from the inner, optically thick core, the surface density can also be derived.

Fine details also have some influence. For example, the disk thickness has a non-negligible impact, resulting in a difference of inclinations between the near and far side of the disk surface and, thus, a modification of the projected velocity. This substantially distorts the emission regions as a function of velocity. Thus, the scale height and even the flaring index can be constrained even with what appears to be insufficient resolution (i.e., resolution coarser than the scale height) [102]. Similarly, one can envisage constraining the vertical location of molecules above the disk mid-plane; molecules appearing only above 1 or 2 scale heights would emit like two superposed disks of slightly different inclinations (10 to 20 degrees). While this can be difficult to constrain with current mm arrays, ALMA largely has the potential for such studies [110].

Fig. 26 Typical sensitivity for line (and continuum) observations of the existing arrays at 0.5″ resolution. The expected brightness of transitions from the CO isotopologues in "normal" disks is also indicated

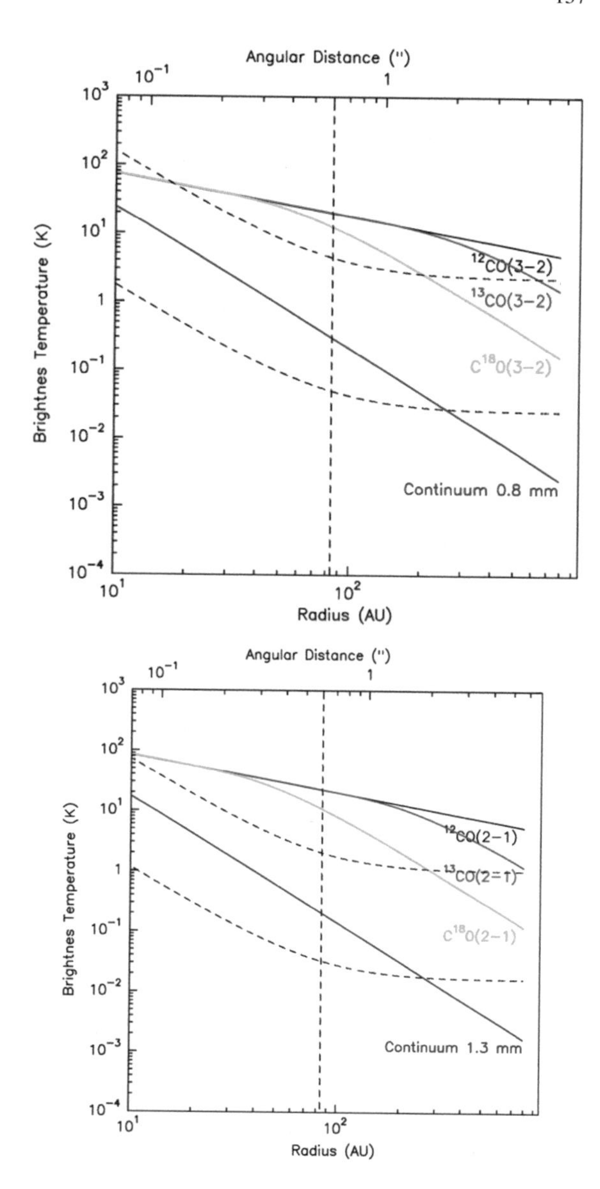

Current Results and ALMA Prospects

The properties of molecular emission have been used to constrain the physical parameters of protoplanetary disks. An essential limitation is the sensitivity of current arrays; an example is given for 0.5″ resolution in Fig. 26.

The typical sensitivity is of order 1 K, so optically thick lines can be easily detected; however, optically thin lines, such as those of $C^{18}O$, become difficult to observe beyond 100 AU or so. Also, only the external regions ($r > 10$–30 AU) give sufficient flux to be observable. Accordingly, the spatial dynamic range of current studies is limited.

Another effect of the limited signal to noise of current studies is linked to the relatively limited UV coverage provided by existing mm arrays. This leads to substantial side lobes that require nonlinear deconvolution to produce understandable images. However, noise is amplified during the deconvolution process. In particular, the reconstruction of extended structures depends on the effective signal to noise. Obtaining a deconvolved image of the velocity-integrated emission is particularly challenging. Integrating prior to deconvolution results in using detection bandwidth that is much larger than the local line width, which is not optimum for signal to noise. On the other hand, deconvolving prior to averaging makes it difficult to reconstruct extended structures. In practice, there is no ideal solution. Furthermore, removing the continuum emission must be done with some care (but not forgotten!).

In practice, until ALMA is fully operational, one should be cautious about analysis that only uses images. A (partial) solution to the problem is to use UV plane analysis, which avoids the nonlinear deconvolution stage. The generic principle always follows the same path: (i) define a (parametric) disk model (e.g., a truncated power-law disk); (ii) if at LTE, compute the 3-D data cubes (x,y,v) with a ray-tracing tool (non-LTE cases require us to solve the coupled problem of statistical equilibrium and radiative transfer to compute the source function); (iii) compute the theoretical visibilities of these sets of 2-D images at the same (u,v) points as the observed visibility; and (iv) compare these with observations with a χ^2 criterium, using the known weight of each observation. Generic tools can locate the best χ^2 as a function of the model parameters.

Astronomers who are unfamiliar with mm radio astronomy are sometimes surprised by the fact that the best-fit criterium in this process uses a nonreduced χ^2. This is because each visibility carries very little information about the disk model, which is compensated by the large number of observed visibilities. The nonreduced χ^2 is

$$\chi^2 = \sum_i (\mathrm{Re}(\mathrm{Model}_i) - \mathrm{Re}(\mathrm{Obs}_i))^2 + (\mathrm{Im}(\mathrm{Model}_i) - \mathrm{Im}(\mathrm{Obs}_i))^2/\mathrm{Noise}_i^2, \quad (40)$$

where Re and Im indicate the real and imaginary parts of the visibility. The observed visibility $\mathrm{Obs_i}$ is the true source visibility Source_i plus a random noise of expectation Noise_i. For most visibilities, we have (sometimes very) low signal to noise, i.e.,

$$(\mathrm{Re}(\mathrm{Source}_i)^2) + (\mathrm{Im}(\mathrm{Source}_i))^2 \ll \mathrm{Noise}_i^2. \quad (41)$$

Under these circumstances, even a null model leads to a value of χ^2, that is just slightly larger than the number of observations, i.e., a reduced χ^2 that would deviate only very marginally from 1. In fact, the deviations of the reduced χ^2 from its expected value of 1 only indicates how accurately the noise has been predicted by the radiometric equation.

As a result, it is impossible to tell from the χ^2 whether a fit is good or not. The only possible solution is to make an image of the fit residuals, Model − Obs, and to verify that the image only contains noise, especially where the Model or Obs are non-zero.

The use of χ^2 model fitting started with the earlier study of [60] and has become a standard approach since then. Different minimization techniques have been used (simple grid search, modified Levenberg–Marquardt, or even Monte Carlo Markov Chains). Error bar estimates can usually be provided, but their interpretation is not always straightforward because of some coupling between parameters. The most important couplings are discussed by [31, 102].

In practice, the sensitivity limitations have restricted the studies to the brightest lines and most parametric studies assumed truncated power-law models. Beyond the answer to the simplest question (Do disks exist? Yes, they do.), interferometric and single-dish mm observations have revealed a number of essential properties of protoplanetary disks. ^{12}CO studies showed that disks can be big (several 100 AU, up to 1300 AU). In fact, the observed sample is highly biased toward the largest disks since the integrated line flux of an optically thick line goes as $R_{\rm out}^{2-q}$. Disks do rotate in Keplerian rotation: [115] derived exponents $v \approx 0.50 \pm 0.02$ for a few disks. This places an upper limit on the (outer) disk mass, which is never large enough to affect the rotation curve. Disks exhibit radial temperature gradients; [60] derived $q = 0.65 \pm 0.05$ for DM Tau. As expected, disks around HAe stars are warmer than disks around T Tauri stars [100, 101, 115], but there is no direct correlation with spectral type. Also, turbulence must be subsonic since the local line widths are around 0.15 km s^{-1} or less [102].

Beyond these rather expected results, a well-established fact is that CO is under-abundant compared to disk masses derived from dust emission; see Sect. 7.3. The depletion factor is typically estimated to be 10, e.g., [31], but AB Aur, an A0 HAe star surrounded by a large, warm disk, shows no obvious CO depletion [101].

This suggests that, as in protostellar cores, condensation on dust grains is responsible for the depletion because in T Tauri disks the temperature is below the freezeout temperature of CO ($\simeq$17 K) over a large portion of the disk. The densities are so high in disks that depletion should occur on a very short timescale. One difficulty with this interpretation resides in the fact that the apparent brightness of (optically thick) CO is often well below 15 K.

Much larger depletions ($\simeq$100) have been observed in other disks: BP Tau [41], CQ Tau, and MWC 758 [22]. Yet these three disks are warm, $T_k > 30$ to 50 K, making depletion an unlikely mechanism. [22] showed that enhanced photodissociation due to grain growth (because the opacity goes as $\sqrt(a_{\rm max})$ for a size distribution with exponent 3.5) could lower the CO column density for stars with high UV flux, such as HAe stars. However, these stars would produce strong C and C$^+$ emission, which is not detected [23]. Low gas-to-dust ratios may have to be invoked instead.

CO has not only been detected around weak-line T Tauri stars (wTTs), but also in a "transition" object between wTTs and classical T Tauri stars, i.e, V 836 Tau [42]. This may suggest that outer disks (traced by CO) disappear on the same timescale as inner disks (traced by IR excess and accretion signatures), but the statistics are still practically inexistent.

7.3 Continuum Emission

Continuum emission from dust in circumstellar disks may a priori appear to be a simpler problem to understand than line emission. At mm wavelengths, the dust is largely optically thin and the radius of the optically thick regions is not expected to exceed a few to a few tens of AU (for the most massive disks). Furthermore, the Raleigh-Jeans approximation is in general good, as expected dust temperatures are in the range 10–30 K near 100 AU. The emitted flux should thus scale linearly with disk mass and disk temperature. Since very wide bandwidths can be used to detect continuum emission, mm arrays are a priori ideal tools to study the shape of surface density dependence upon radius.

For a power-law radial distribution, the brightness temperature decreases as $r^{-(p+q)}$ in the outer, optically thin, regions, but only as $r^{-}q$ in the inner optically thick regions. With sufficiently high angular resolution, one could thus simultaneously measure the dust temperature (T, q) and the surface density Σ, p (assuming dust emissivity is known and does not depend on radius r).

Unfortunately, the typical arcsec-like angular resolution reached in most observations is not sufficient to well resolve even the closest disks (60–140 pc, in the TW Hydrae Association, ρ Oph or Taurus regions). This results in several complications.

To illustrate some of the issues, consider the typical power-law case shown in Fig. 26. On one end of the power law, the brightness sensitivity limits the imageable area to $r < 200$ AU or so, leaving only a few independent radii. On the other end, the inner thick core contributes to a substantial flux, but remains unresolved. Viewed face-on, the apparent size of such a disk would scale linearly with the effective angular resolution of the observation; since sensitivity issues make the disk edge irrelevant, we are left with a power law with no characteristic size. [39] actually used this property to demonstrate that the exponent p must have been ≤ 1.5 in most disks because the disks appeared marginally resolved at about $1''$ resolution.

If viewed edge-on, even the disk orientation could be significantly biased when observed with an elliptical beam. Along the major axis, the apparent size depends on the effective resolution, while along the minor axis, the surface density falls as a Gaussian (in the isothermal approximation) with well-defined characteristic size. This property largely explains the different disk orientations cited in the earliest studies [39, 77]. Only the most recent, highest angular resolution studies provide reliable dust disk orientations [2, 63, 74].

A second complication is that the dust temperature cannot be constrained when the optically thick inner core is not resolved. Most studies have alleviated the problem using temperatures derived from the IR SED. However, strong vertical temperature gradients exist and the large grains responsible for the mm emission are expected to be colder than the small grains sitting 1 or 2 scale heights above the disk plane providing the (optically thick) IR emission. At least a two-layer (two temperature) disk model, such as that of [25], is required to properly estimate the effective temperature [73, 77]. Without using this approach, the derived masses can be severely affected. Indeed, [64] find a much larger mass for the disk of MWC 480 than [101], who derive the temperature from the apparent brightness of the optically thick core. Another potential effect is that large grains may have lower physical temperatures than smaller grains because of their different radiation properties.

A second complication resulting from insufficient sensitivity is that the outer radius is poorly constrained. In the power-law model, the surface density exponent p may be derived with sufficient precision and the outer radius is adjusted to match the total flux (i.e., the visibilities at short spacings). There is thus substantial coupling between R_{out} and p.

In fact, the combination of limited angular resolution and sensitivity makes the characterization of the true shape of the surface density profile difficult. One of the surprising results of disks observations analyzed by power-law models is the much smaller outer radius derived from dust compared to that derived from other molecules. [71] suggested that both properties could be reconciled with a different surface density profile, namely, an exponentially attenuated power law. Such density profiles are self-similar solutions to the viscous evolution when the viscosity is a (constant in time) power law of radius. Using a few examples (TW Hya, MWC 275, GM Aur), [71] showed that a single disk model could explain the apparent disk sizes in CO and mm dust emission. This can be easily understood as CO is essentially optically thick: CO traces the temperature profile, while the dust traces the surface density. However, it remains unclear whether the viscous model provides adequate explanation for the outer radii derived from more optically thin molecular tracers, in particular, the ^{13}CO lines.

7.4 Model Dependencies

Assumptions about the underlying model can have strong consequences for the derivation of the disk parameters. For example, assume that the disk surface density follows an exponentially attenuated power law, as predicted by viscous evolution models (see, e.g., [71]), and that the emission is optically thin. Assume a constant temperature for simplicity. If emission from such a disk is analyzed under the assumption of a single truncated power law, the power law needs to be steep to represent the outer regions and the flattening in the inner region may be misinterpreted as the result of an optically thick core. The low brightness of the core would be misinterpreted as a cold medium, while this low brightness is in reality caused by optically thin emission. This degeneracy can be (partly) removed by multifrequency observations. The radius at which $\tau = 1$ is frequency dependent because the dust (and also spectral line) emissivity depends on the frequency. Thus, the flattening due to optical depth effects would appear at different radii in the power-law hypothesis. On the other hand, changes in slope of the surface density result in a frequency independent effect.

Another example of model dependent results is given by the assumption of hydrostatic equilibrium. This assumption couples the flaring of the disk to its physical temperature. Hence, a model assuming hydrostatic equilibrium could be forced to adjust the temperature in response to an apparent flaring, rather than simply to adjust the apparent brightness. As a result, this (perhaps incorrect) assumption about the temperature could affect the inclination (for an optically thick transition) or the column density (for the optically thin case). Some of these degeneracies are discussed by [102].

8 High-Mass Stars

The formation of high-mass stars is a specially challenging problem for which mm/sub-mm interferometers can play a major role. Because of the rarity of high-mass stars, the nearest high-mass star formation regions are further away from the sun compared to low-mass star-forming regions. The nearest region is the Orion nebula, 410 pc [90], but it may be unusual in several respects. More typical regions (the Cygnus area, including DR21, and the W3(OH) core, e.g., [125]) are 1.7–2 kpc away. Furthermore, high-mass stars never form in isolation. They are embedded in massive molecular clouds, where the dust extinction is important. Accordingly, observations with single-dish telescopes remain very confused because they lack the resolving power to separate the observed regions into individual objects. Thus, mm/sub-mm arrays are key tools to unveil the structure and properties of high-mass star-forming regions.

8.1 Toroids and Outflows

A major question is whether high-mass stars form like low-mass objects with circumstellar disks to regulate the angular momentum. High-velocity outflows are ubiquitous in high-mass stars, however, their origin is often unclear because of the multiplicity of pre-main sequence stars around the most massive star. Rotating structures have been discovered in many high-mass star-forming regions, suggesting that indeed disks may play an essential role. However, these structures are massive toroids, not simple circumstellar disks [12]. The dynamical mass derived from the rotation velocity is, in general, substantially larger than the stellar mass inferred from the object luminosity. These toroids are also relatively thick ($H/R \simeq 0.5$).

In discovering the rotating structures, a key feature of mm arrays has been used: Their ability to provide *super-resolution* for the kinematical information. The relative positions of the emission centroids as a function of velocity can be determined with much higher precision than the synthesized beam. The reason for such a capability

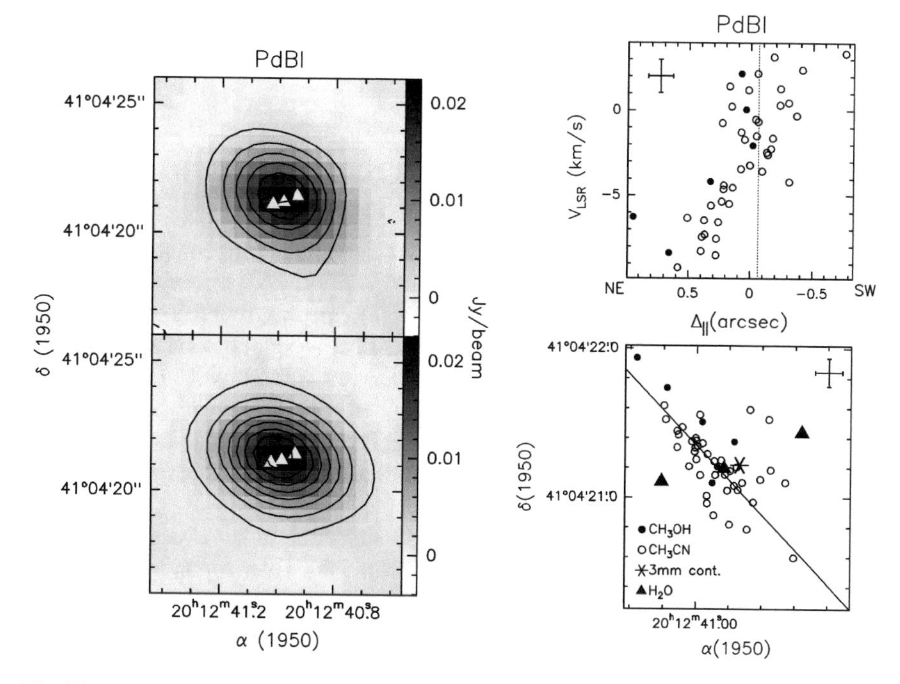

Fig. 27 Example of the use of super-resolution to infer the presence of rotating structures in high-mass star formation regions (from Cesaroni et al. 1997 [19])

is that relative positions in adjacent velocity channels depend on the relative instrumental phases of the array, not on the atmospheric phase. Accordingly, the relative position accuracy from channel to channel is limited (besides the usual signal-to-noise limitation) by the bandpass (phase-only) calibration accuracy. In most arrays, this is on the order of a few degrees of phase, i.e., 1/100th of the synthesized beam. An example is given in Fig. 27. Some arrays even have specific calibration devices to ensure high bandpass calibration accuracy. The IRAM Plateau de Bure array includes a built-in noise source that monitors the time variations of the correlator phases. This system results in typical bandpass phase error below a fraction of a degree. The super-resolution accuracy, however, is also limited by the precision of the delay tracking system since a delay error introduces a constant phase slope as a function of frequency.

The existence of rotating toroids provides indirect information about the origin of the high-velocity molecular outflows. In most cases, the outflow is nearly symmetric about the toroid center and its direction is aligned with the rotation axis of the toroid [12]. This suggests that the outflows indeed originate from the massive central object rather than from one or more lower mass stars of their surrounding. However, outflows also exhibit even wider complexity than those from low-mass stars. Multiple outflows are more the rule than the exception [14] and evidence for precessing jets is frequent [116]. Quadrupolar outflows are also frequent in intermediate-mass objects

[57]. These properties are likely to be related to the higher multiplicity observed for more massive stars.

8.2 Hot Cores

The huge luminosity of high-mass (proto-)stars raises the temperature of the surroundings to much higher values than those found in low-mass star formation regions. *Hot cores*, with temperatures in excess of 100–300 K, are thus found. These hot cores often correspond to the rotating toroids, but are sometimes not obviously associated with the most luminous object. For example, the hot core in W3(OH) is actually not associated with the ultracompact HII region, but with an embedded object a few arcsec away [125]. The high temperature of the hot cores results in evaporation of the grain mantles formed in the cold molecular cloud phase and in peculiar "warm", time dependent chemistry [104].

These hot cores thus exhibit strong line emission from molecules, which would remain trapped in ices in cold regions, such as ammonia, water vapor, methanol, etc. They also display line emission from even more complex molecules and are ideal regions to study the molecular complexity. Indeed, apart from the unusual warm cloud Sgr B2 in the direction of the galactic center, most complex molecules have been discovered in high-mass star formation regions, and more specifically the nearest of these regions, Orion.

The molecular complexity offers advantages and drawbacks. Because heavier molecules have lower rotational constants, these molecules provide a much larger number of transitions within a given frequency range than the usual diatomic molecules used as tracers for low-temperature gas. This can be used to provide accurate thermometers. *Rotation diagrams*, also called Boltzmann plots, which display the line intensity as a function of lower level energy, allow us to derive the effective rotation temperature (if it applies to all observed levels) and the molecule column density. The method derives from the brightness temperature expression

$$\Delta T_\mathrm{b} = J\nu(T_\mathrm{ex})(1 - e^{-\tau}) = J\nu(T_\mathrm{ex})\tau\frac{1 - e^{-\tau}}{\tau}, \tag{42}$$

where N_u is the column density in the upper level per Zeeman sublevel. By definition of the opacity

$$J\nu(T_\mathrm{ex})\tau = \frac{hc^3}{8\pi k\nu^2}\frac{\rho}{\Delta V}g_u A_{ul} N_u, \tag{43}$$

where ρ is a shape factor relating the integrated line intensity W to the peak brightness ΔT_b and line width ΔV, i.e.,

$$W = \int \Delta T_\mathrm{b} dv = \Delta T_\mathrm{b} \Delta V/\rho. \tag{44}$$

From Eqs. 42–43, we obtain

$$\Delta T_{\mathrm{b}} = \frac{hc^3 A_{ul}}{8\pi k \nu^2} \frac{\rho}{\Delta V} g_u N_u \frac{1 - e^{-\tau}}{\tau} \tag{45}$$

and introducing

$$\gamma_u = \frac{8\pi k \nu^2}{hc^3 g_u A_{ul}}, \tag{46}$$

which can be written as

$$\Delta T_{\mathrm{b}} = \frac{1}{\gamma_u} \frac{\rho}{\Delta V} N_u \frac{1 - e^{-\tau}}{\tau}. \tag{47}$$

For purely optically thin lines, $\frac{1-e^{-\tau}}{\tau} = 1$, so

$$W = \Delta T_{\mathrm{b}} \Delta V / \rho = \frac{N_u}{\gamma_u}. \tag{48}$$

If we assume all levels are described by a Boltzmann distribution at temperature T_r,

$$N_u = N/Z \exp(-E_u/(kT_r)), \tag{49}$$

where Z is the partition function, and in log form,

$$\log(N_u) = \log(N) - \log(Z) - E_u/(kT_r) \tag{50}$$

which, according to Eq. 48, translates into

$$\log(W) = \log(N) - \log(Z) - E_u/(kT_r). \tag{51}$$

Thus, the slope of $\log(W)$ gives the rotation temperature T_r, which in turn allows the derivation of the partition function Z, and the column density can then be derived from the intercept at $E_u = 0$. A more thorough description is given by [51]. Although convenient and widely used, this method has several pitfalls. One obvious pitfall is the optical depth issue. The method is only valid if all lines are optically thin. Opacity effects can mimic a temperature dependence as a function of line energy. The temperature dependence resulting from opacity effects can be recognized for linear molecules, but may be much more complex for molecules with a more complicated spectrum such as CH_3OH. A second effect is source size correction: Observations at different frequencies are often carried out with different beam sizes. Converting the apparent source brightness to an intrinsic value requires correction for the beam filling factor prior to comparison in a rotation diagram.

An often forgotten issue is that the whole derivation is based on homogeneous line formation assuming only microturbulent line width. Unfortunately, temperature gradients, which are unavoidable in internally heated hot cores, can profoundly affect the rotation diagram. The impact of velocity gradients (whose existence is proven by the images) is more debatable. For purely optically thin lines, it should not matter because all photons escape and contribute to integrated line intensity. However, in case of partially optically thick emission, the result depends on the details of the velocity field and on the effective coherence length of the turbulence compared to the $\tau = 1$ scale length. Therefore, attempts to correct for the opacity effects may fail if the incorrect source model is used.

Another effect to consider is contamination by the continuum (mostly thermal dust emission, although free-free emission should not be ignored for ultracompact HII regions). At high frequencies, the dust opacity can become non-negligible, and improper subtraction of the continuum emission then affects the derived line intensity.

CH_3CN and CH_3C_2H are especially good thermometers because they exhibit rotation lines from the same J number, but different K quantum numbers (and thus different E_u) at very nearby frequencies. Opacity effects are easier to identify with these molecules (as they affect low K transitions more than high K transitions).

8.3 Molecular Complexity

Since they are rich in molecules, hot cores are the ideal sources toward which searches for new molecules have been carried out. Orion A and W51 are among the favorite candidates. However, the density of lines per unit frequency becomes so important that *line confusion* is a limiting factor to identify new molecules. New (i.e., more massive or more complex) molecules have a more complex spectrum than those currently identified. Their spectra thus have a larger density of lines as a function of frequency compared to smaller molecules, but the line intensities are smaller (as the same energy input can be dissipated over more transitions). Line confusion becomes a severely limiting factor to identify new species. For example, consider the spectrum ethyl-methyl-ether (EME, $CH_3OC_2H_5$). Tentative detection was reported in Orion KL by [24], but it is based on a single transition. The possibility that emission at this single frequency comes from another molecule is not small because a fair fraction of lines in Orion KL remain unidentified. [46] searched for several transitions of the same molecule (actually, the trans conformer, which has lower energy than the cis conformer, and is thus expected to be more abundant). The expected spectrum of EME at 130 K is given in Fig. 28 in the mm/sub-mm domain. The most intense line in the mm domain, at 237 GHz, is unfortunately blended with that of another molecule. Clearly, a positive detection of EME would require matching all (nonblended) lines with their respective intensity ratios. Unfortunately, the expected signal is masked by that of other molecules, as shown for example around 150 GHz for W51e2 in Fig. 29. The synthetic spectrum, obtained with an adjustable source model and precise line frequencies for all known molecules, matches the main features of the observed spectrum reasonably well. However, some strong lines remain unidentified and others appear to have incorrect intensities, perhaps due to

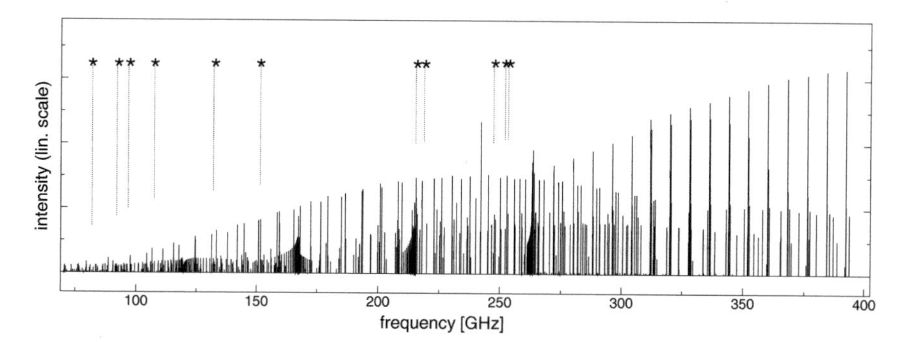

Fig. 28 Theoretical spectrum of Ethyl-Methyl-Ether in the mm/sub-mm domain (from Fuchs et al. 2005 [46])

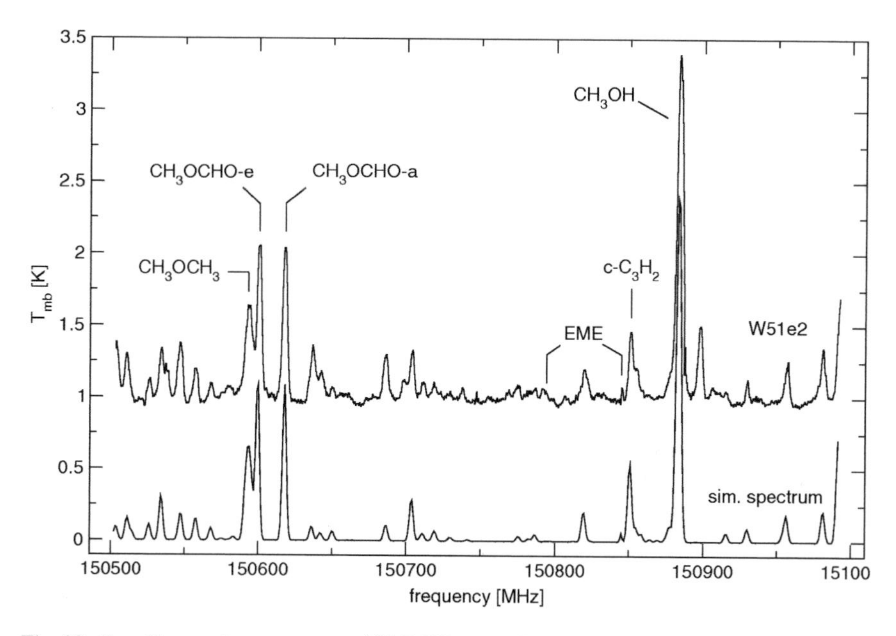

Fig. 29 Top: Observed spectrum near 150.7 GHz toward the massive star-forming region W51N. Bottom: Synthetic spectrum including all known molecules, showing the level of confusion when looking for faint species like EME, and revealing, by comparison, some unidentified lines in the observed spectrum (from Fuchs et al. 2005 [46])

blending with an unknown molecule. The signal from EME itself is very weak, if present at all.

This example is a good illustration of the difficulty in identifying new molecules, despite a sophisticated global model fitting approach to best represent known molecules. Accurate line frequencies are mandatory to these studies and their acquisition is a major challenge for laboratory spectroscopy. Among these, isotopic substitutes are essential. Until their recent laboratory determination, frequencies from rel-

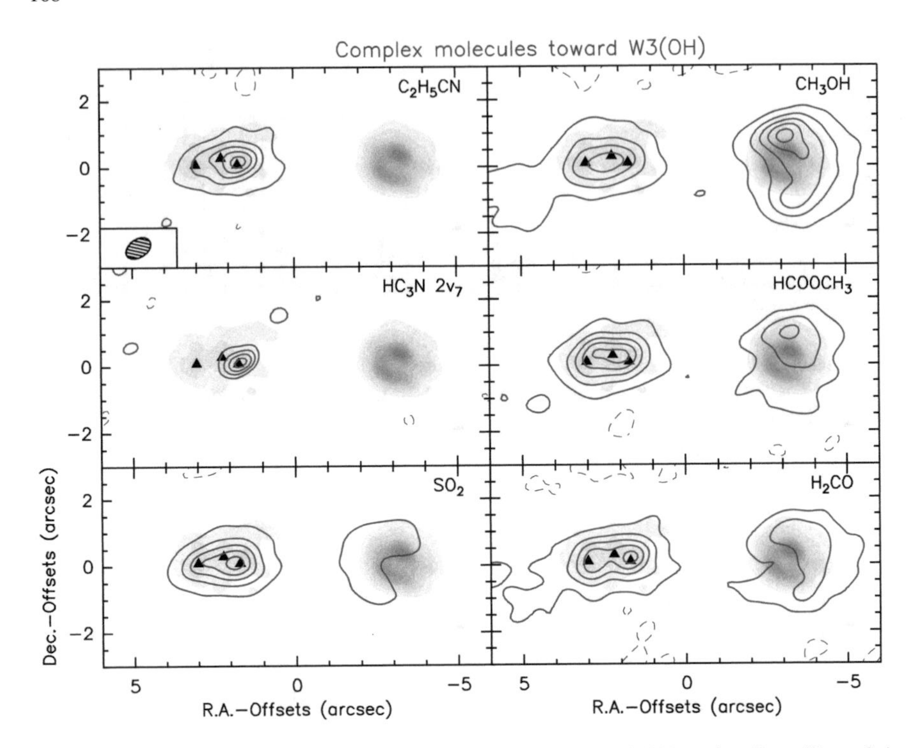

Fig. 30 Chemical differentiation between the two sources in the W3(OH) region (from Wyrowksi et al. 1999 [125])

atively abundant ^{13}C isotopologues of ethyl cyanide ($^{13}CH_3CH_2CN$, $CH_3^{13}CH_2CN$, and $CH_3CH_2^{13}CN$) were not precisely known, and these isotopologues have been shown to match several hundreds of U lines in Orion KL [32]. Because of the high temperatures of hot cores, vibrationally excited states also play a significant role, especially since complex molecules have many vibrational modes and can be relatively "floppy"; see, for example, [33] for methyl formate and ethyl cyanide.

The expected time dependence of the chemistry in high-mass star-forming regions can result in strong chemical differenciations between nearby sources. For example, in W3(OH), O-rich molecules peak toward the UCHII region, while N-bearing molecules peak toward the embedded source (see Fig. 30 from [125]). It is important to take advantage of this distinction. As some molecules only appear in one region, high spatial resolution can reduce the line confusion. An example of such a study is given for the Orion KL region by [52], who show that the complex single-dish spectrum from Orion KL originates in at least four sources with different chemical composition, where only one source is rich in N-bearing molecules. The three other sources, although all rich in O-bearing molecules, have very different spectra.

High angular resolution and complete, accurate line catalogs with isotopic substitutes combined with sophisticated spectral modeling will be required in further attempts to identify more complex molecules, such as the "pre-biotic" glycine or its lower energy isomer methyl carbamate.

Finally, mm frequencies may be more appropriate for such searches than the sub-mm domain. This is because line strengths and, thus, line confusion in rich sources, tend to increase with frequency. Currently, confusion appears to be a more serious limit to that sensitivity.

9 Nonstandard Observations

9.1 *Polarization*

Almost all mm/sub-mm observations ignore the polarization properties of incoming radiation. Yet, although the polarized intensity is usually a very small fraction of the total intensity, the study of polarization can provide essential information about the magnetic field, which is a physical quantity that is otherwise elusive and critical to many processes.

Continuum Polarization

Polarized dust emission originates from aligned spinning (nonspherical) dust, although the alignment mechanism remains a very debated topic [15, 83].

This leads to linearly polarized emission and measuring the polarization vectors can indicate the magnetic field geometry. High-resolution observations at mm wavelengths have been rare not only because of signal weakness, but also because only two arrays (BIMA and the Sub Millimeter Array, SMA) were equipped for polarization observations. Moreover, these arrays did not have dual-polarization receivers, and used a switching scheme to alternate measurements in the four cross-product states required to measure the Stokes parameters. This switching scheme limits the precision that can be obtained since atmospheric changes between alternate measurements can affect the signal. Linear polarization is best measured by correlating circularly polarized signals. Since mm receivers are always equipped with linear feeds to provide the widest possible tuning range, conversion to circular polarization requires the insertion of a quarter wave plate, which restricts the observable frequencies to a narrower range.

Accordingly, measurements are scarce. For low-mass Class 0 sources, SMA observations at 345 GHz by [49] reveal the hour-glass morphology of the magnetic field, which was only previously tentatively identified by lower frequency, lower angular resolution data from BIMA at 230 GHz [48]. Reference [120] also discovered hour-glass morphology in W51 e2 for high-mass star-forming regions, where initial attempts failed to reveal clear morphologies. This further suggests that high-mass stars form through a similar process as low-mass stars with the magnetic field playing an essential role in regulating the infall process.

Marginal detections (at the 3σ level) have been reported in GM Aur and DG Tau from unresolved observations by [119] of more evolved objects, such as protoplanetary disks, in Class II sources. However, spatially resolved stringent upper limits obtained in two sources (HD163296 and TW Hya) by [72] appear inconsistent with theoretical models by [26]. The inconsistency may be due to lower magnetic field, low alignment efficiency, or low grain elongation (perhaps because of grain growth). Positive detections of polarization, spatially resolved and below the 2% level, will be required for further progress in understanding the current discrepancies.

While dust emission can reveal the magnetic field geometry, it is difficult to estimate the magnetic field strength from these observations alone. The Chandrasekhar-Fermi relation [21] relates the field strength to the dispersion of the field directions. However, measurement of the direction dispersion is rather severely noise limited. A signal to noise of 10 (on the polarized signal, i.e., several hundreds on the total intensity for typical polarization levels of few percent) leads to a 1σ error of $6°$ on the direction.

Another specific instrumental problem related to polarization may affect aperture synthesis observations. First, polarized emission is not a positive quantity, thereby limiting the tools that can be used to deconvolve the dirty image. Second, polarized and total intensity distribution have different characteristics scales, and can thus be affected differently by the missing UV (specially the short) spacings. As a result, the polarization fraction can be an ill-defined quantity.

Line Polarization

Line polarization can result from asymmetric radiation fields and/or velocity gradients, which is the so-called Golreich-Kylafis effect [50], whose intensity depends on the magnetic field strength. The expected polarization is linear with polarization levels up to a few percent. Polarization attributed to this effect has been detected from the low J transitions of CO in outflows [48] and in a high-mass star formation region, DR21(OH) [27, 82]. In DR21(OH), the $J = 1 - 0$ and $J = 2 - 1$ transitions have orthogonal polarization, suggesting a combination of velocity gradients and anisotropic radiative excitation. Polarization is quenched by collisions; polarization measured from CO only samples low-density ($\sim 10^3$ cm^{-3}) regions.

Another essential phenomenon is the Zeeman effect. The magnetic moment of the molecule must be large enough for the Zeeman splitting to be significant. In practice, only molecules with unpaired electrons can lead to significant Zeeman splitting. This is only possible for the lowest rotational transitions. In star-forming regions, the best candidates are first CN, then CCH (whose line intensities are in general weaker, however), and possibly also SO and CCS, although their larger molecular weights make them less suitable. The hyperfine structure of the CN and CCH transitions make them particularly useful because the Zeeman splitting is different for each component. Accordingly, the intrinsic signature of the Zeeman splitting can be separated from instrumental effects. An example is given (again for DR21(OH)) in Fig. 31 [29]. Typical field strengths of the order of a mgauss are required for the Zeeman effect to be detectable with CN.

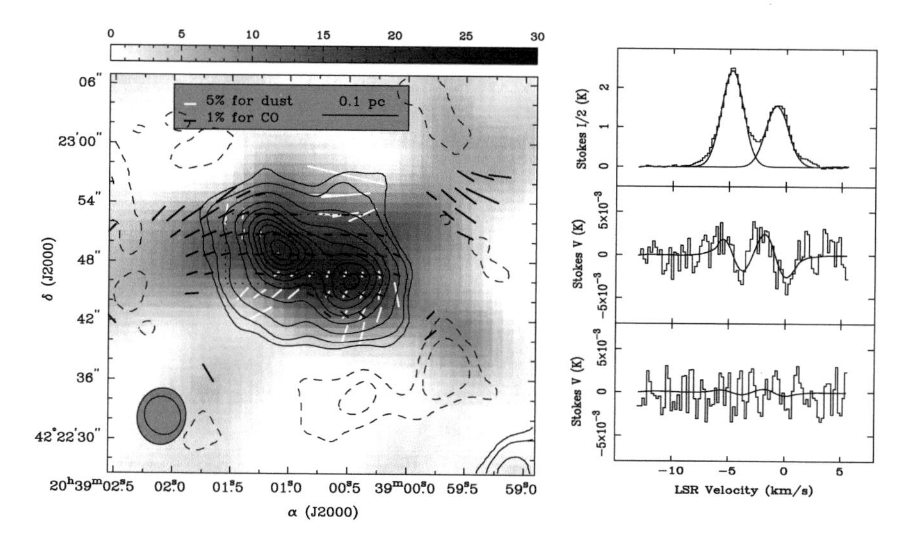

Fig. 31 Detection of the Zeeman effect from the $N = 1 - 0$ line of CN toward DR21(OH) (from Crutcher et al. 1999 [29])

The sensitivity limitation is more critical for line polarization than for dust continuum emission since the detection bandwidth is limited by the line width in the case of line polarization.

9.2 Time Dependent Effects

Millimeter-wave astronomy has mostly dealt with time independent emission. Long-term changes are now becoming accessible since the earliest measurements at high angular resolution now date from more than 20 years ago. Among these, *proper motions* play a significant role.

The astrometric accuracy of mm arrays is sufficiently high that proper motions due to source velocity can be detected up to significant distances. With a 15 year time base, [63] showed that the precision reached for disks around T Tauri stars in the Taurus region is comparable to the Hipparcos measurements, $\simeq$1.5 mas/yr in each direction, and often better than published data based on optical studies spanning the last 30–50 years [36].

Changes in morphology also occur. An obvious case is the proper motion of outflow bullets and the associated expansion motion of the outflow cavities. Measuring these motions enables the retrieval of the outflow inclination through comparison with the projected velocities [47]. All such proper motions should be accounted for when comparing high-resolution studies at different dates.

On shorter timescales, accounting for time variability may become necessary when studying specific objects. An example of this is magnetically active stars.

Intensity variability has been discovered at mm wavelengths in young binary stars, which are on elliptical orbits with small enough perihelion that their magnetosphere can interact. Such variability was reported by [39] for V 773 Tau and also for DQ Tau by [108]. Flux variability at much lower levels may exist in other objects, but cannot be detected by current mm arrays owing to their limited intensity calibration accuracy (typically 10–15%).

10 Conclusions: The Promise of ALMA

ALMA will unavoidably bring major improvements to our knowledge of the star formation processes, both for low- and high-mass stars. Compared to existing arrays in the pre-ALMA era, ALMA brings major improvements in several areas. These improvements are summarized below:

- Sensitivity improvement. ALMA will improve sensitivity by a factor $5-10$ (in the mm domain, compared to the IRAM or CARMA arrays) to 100 or more (in the sub-mm domain, compared to the SMA).
- Image fidelity. ALMA will be the first instrument able to produce images that deviate less than $1-2\%$ from the true sky brightness distribution (a fidelity of 50–100). For comparison, images obtained by current arrays rarely exceed fidelities of 5.
- Imaging speed. Imaging speed is obtained with a combination of the sensitivity gain (a gain S in sensitivity implies a gain in observing time of S^2 to reach the same result) and of instantaneous UV coverage, which makes ALMA no longer dependent on Earth rotation for aperture synthesis. Imaging speed is essential to allow surveys of a substantial number of sources, which would uncover general properties
- Wide-field imaging capabilities. The overall ALMA concept involves the ALMA Compact Array and a single-dish mode to properly recover all angular scales from the synthesized beam up to the imaged field size.
- High spectral resolution. Most arrays have been limited to spectral resolutions of order 0.05–0.1 km s^{-1}, either intrinsically (by the correlator hardware) or by insufficient sensitivity. ALMA will be able to go down to 10 m s^{-1} or less, which can be essential to detect subtle motions or to probe subsonic turbulence.
- Built-in full-polarization capabilities. Polarization studies have been performed with current arrays, but via dedicated hardware and, thus, for a limited sample of sources or frequencies. In practice, this occurs only for continuum emission or very bright lines like CO or SiO maser emission. ALMA can do that systematically, although at the (moderate) expense of a factor 2 in spectral resolution and/or total bandwidth.
- Frequency coverage. ALMA will cover high frequency bands at high angular resolution for the first time.

- Instantaneous frequency coverage and multiplex advantage. The large instantaneous bandwidth of ALMA receivers (several GHz), combined with the versatility of its correlator, will allow observations of several spectral lines at once with identical calibration characteristics, greatly improving the comparisons between them.
- Calibration accuracy. Image fidelity can be divided into two independent aspects: the morphology fidelity, i.e., the ability to correctly represent relative brightness changes, and the absolute flux scale accuracy. While current arrays were rarely exceeding 10–15% flux accuracy, especially at the high end of their frequency coverage, ALMA has the potential to reach calibration accuracies of order 1–3%. This will considerably improve the ability of astronomers to compare images at different frequencies, for example, as required in multiline studies to constrain excitation conditions, and thus temperature and densities.
- Astrometric accuracy. This kind of accuracy will be considerably increased from those of current arrays by the fast-switching capability of ALMA for phase calibration. The gain in positional accuracy comes from the reduced atmospheric phase residual due to the fast timescale and from the reduced angular distance between source and calibrator.
- Angular resolution. With synthesized beams, angular resolution is an order of magnitude smaller than the best resolution previously obtained.

However, these exquisite capabilities cannot all be obtained together: A survey of 100 sources, at $0.05''$ resolution, to detect a very optically thin line, e.g., $HC^{15}N$ $J = 3 - 2$ in protoplanetary disks to study $^{14}N/^{15}N$ fractionation, will still be beyond ALMA capabilities. And also, while many 8 m class optical telescopes exist, there is (unfortunately) only one ALMA.

Thus, although ALMA has been designed to be useable by nonexperts, it remains essential to properly plan ALMA observations to reach any specific scientific goal. Here are some guidelines that may be helpful for those with no prior experience in mm (or sub-mm) observations. There are two separate, but not independent (see below), parameters to define an ALMA observation: the angular resolution and frequency setup, i.e., observing frequency and spectral resolution. More complex observing strategies, such as multitransition and wide-field imaging, are beyond the scope of the simple guidelines given below, which only cover simple cases.

The principle issue in proposal preparation is always sensitivity, which can be estimated from the relatively simple noise equation. However, we must first distinguish between detection experiments and imaging studies.

Detection Experiments

Contrary to the northern sky, where the IRAM 30 m and JCMT 15 m or the CSO 10 m telescope were well matched in sensitivity for unresolved studies of sources followed up by the mm (IRAM, CARMA) and sub-mm (SMA) arrays, there is no single-dish radio telescope in this frequency domain large enough in the southern sky. Source searches may often need to be performed by ALMA itself, and detection experiments have to be performed in a compact configuration. As soon as one starts resolving some substructure, the signal to noise decreases because some visibilities are smaller than others. Also, more compact configurations are less affected by atmospheric effects (phase noise). So the first question is "What is the expected maximum (angular) size of the studied source(s)?". The angular resolution should be coarser than this angular size. The only exception to this rule is spatial confusion; the need to separate the source of interest from extended emission or other nearby sources may impose a higher angular resolution. For unresolved sources, the flux sensitivity is

$$\Delta S_\nu = \frac{J T_{\text{sys}}}{\eta \sqrt{N(N-1)\Delta\nu \Delta t}}, \tag{52}$$

where T_{sys} is the system temperature, N the number of antennas, $\Delta\nu$ the total bandwidth (including both polarization for unpolarized signals), and Δt the integration time. The value J is the conversion factor between antenna temperature and flux density for a single antenna, which is about 30 Jy/K for ALMA antennas at mm wavelengths (12 m diameter) and up to 45 Jy/K at the highest frequencies. The value η is an efficiency factor of order 1, which is smaller at the highest frequencies.

For pure continuum observations, the frequency is defined by atmospheric transparency. It is best to select an optimal transparency region, or more precisely, the lowest system temperature appropriately weighted by the spectral index of the emission α. As for dust emission, $\alpha \sim 2 - 3$, in general, source searches are optimally carried out near 350 GHz. A strong caveat concerning this choice is the size of the field of view over which the search should be performed. If the search requires several overlapping fields, the optimal frequency is around 270 GHz because the number of required fields increases with frequency.

If the science objective is to study the dependency of the spectral index α, the other frequency(ies) will need to be as much separated as possible from the first frequency, while still keeping sufficient signal-to-noise. Accurate flux calibration will matter in this case, and it will in general be preferable to select frequencies below 350 GHz for such studies.

When attempting to detect a molecule, the choice of transition can be guided by the general properties of line emission as a function of energy level (see Sect. 3). However, other considerations apply. A molecule may be selected because it is a temperature tracer (e.g., CH_3CN or CH_3C_2H) or a density tracer (e.g., CS). Bright lines, such as those of CO, may be preferred, however, one should be aware of the possible confusion with the surrounding molecular clouds, which is unavoidable in star formation studies. This may justify selecting optically thinner tracers.

The bandwidth coverage should be sufficient to properly identify the continuum level. As lines are Doppler broadened, the expected velocity dispersion in the source can be used to define this bandwidth. The spectral resolution should be high enough to resolve the line. It is always possible to degrade the spectral resolution afterward, and observing with higher than a priori required resolution can unveil some unexpected property (e.g., self-absorption in a narrow range of velocity). The finest spectral resolution to be used is in practice often set by software limitations; the total number of spectral channels available for a given experiment is limited by the acceptable data rate in ALMA.

Imaging Studies

The first question to address in the case of imaging is: What brightness sensitivity is needed for the scientific goal? Angular resolution comes next because it is essential to first detect a source before resolving it. This derives from the following noise equation for resolved sources:

$$\Delta T_{\mathrm{b}} \approx \frac{T_{\mathrm{sys}}}{\eta N \sqrt{\Delta \nu \Delta t}} \left(\frac{\theta_p}{\theta_s} \right)^2 , \tag{53}$$

given here for a single field of view. The parameter η is an efficiency factor, typically of order 0.5; θ_p is the primary beam size; and θ_s the synthesized beam size, where $\theta_p/\theta_s \simeq B_{\mathrm{max}}/D$ is the ratio of longest baseline to the antenna diameter.

The system temperature is governed by the observed frequency, as the primary beam size θ_p. A good rule of thumb for ALMA is about 0.6 K of T_{sys} per GHz of observing frequency below 350 GHz and about 1.5 K per GHz above 370 GHz. The integration bandwidth is either 8 or 16 GHz (for the continuum) or dictated by Doppler motions in the source (turbulent or thermal line widths). Thus, the only free parameters are angular resolution θ_s and integration time Δt.

The brightness sensitivity in a given time Δt goes as $(1/\theta_s)^2$. Conversely, as the noise only decreases as $1/\Delta t$, the integration time to reach a constant brightness sensitivity goes as $(1/\theta_s)^4$. This very steep dependency makes the brightness temperature the key ingredient to estimate the feasibility of an observation and to derive the compromise between angular resolution and integration time.

For spectral line observations, once the molecular transition to be observed has been selected, in general it is a relatively simple matter to evaluate the expected brightness temperature. The (expected) kinetic temperature of the observed source can be reasonably guessed (often easily within 20%), so it is the line opacity that defines the required brightness. Simple models can help to evaluate a range for this opacity, or more specifically, to convert between opacity and molecular content.

When there is no specific reason to select one molecule or molecular transition, it can be wise to take advantage of the wide bandwidth receivers and flexible correlator to observe more than one line at once. Among the most interesting combinations are the capabilities to observe CO, ^{13}CO, and C^{18}O simultaneously in Band 3 ($J = 1 - 0$ lines) or in Band 6 ($J = 2 - 1$) and CO and ^{13}CO $J = 3 - 2$ in Band 7, but there are many other possibilities involving simple molecules like CN, CS, HCN, and HCO^+. Molecules with hyperfine structures, like CN or N_2H^+, can also sample a large range of opacities with a single transition. Limitations in the average data rate, however, may impose a compromise on the total number of lines that can be covered simultaneously at high spectral resolution.

The problem is often more complex for continuum emission, in fact, because we have no good prior knowledge of the dust properties; an idea about the source size and total flux is required to predict a brightness temperature. However, ALMA is often much more sensitive in continuum than for spectral lines due to its wide bandwidth, so continuum can often be obtained as a serendipitous result from line observations.

References

1. Agladze, N.I., et al.: Nature **372**, 243 (1994)
2. Andrews, S.M., Williams, J.P.: Ap. J. **659**, 705 (2007)
3. Arce, H.G., Sargent, A.I.: Ap. J. **646**, 1070 (2006)
4. Bachiller, R., Perez Gutierrez, M.: Ap. J. **487**, L93 (1997)
5. Bachiller, R., et al.: Astron. Astrophys. **173**, 324 (1987)
6. Bachiller, R., et al.: Astron. Astrophys. **299**, 857 (1995)
7. Bachiller, R., et al.: Astron. Astrophys. **372**, 899 (2001)
8. Bacmann, A., et al.: Ap. J. **585**, L55 (2003)
9. Beckwith, S.V.W., et al.: AJ. **99**, 924 (1990)
10. Beckwith, S.V.W., Sargent, A.I.: Ap. J. **402**, 280 (1993)
11. Belloche, A., André, P.: Astron. Astrophys. **419**, L35 (2004)
12. Beltrán, M.T., et al.: Ap. J. **601**, L187 (2004)
13. Benedettini, M., et al.: MNRAS **381**, 1127 (2007)
14. Beuther, H., et al.: Astron. Astrophys. **387**, 931 (2002)
15. Bethell, T.J., et al.: Ap. J. **663**, 1055 (2007)
16. Bösch, M.A.: Phys. Rev. **40**, L879 (1978)
17. Burrows, C.J., et al.: Ap. J. **473**, 437 (1996)
18. Ceccarelli, C., et al.: Astron. Astrophys. **338**, L43 (1998)
19. Cesaroni, R., et al.: Astron. Astrophys. **325**, 725 (1997)
20. Chandler, C.J., Richer, J.S.: Ap. J. **555**, 139 (2001)
21. Chandrasekhar, S., Fermi, E.: Ap. J. **118**, 113 (1953)
22. Chapillon, E., et al.: Astron. Astrophys. **488**, 565 (2008)
23. Chapillon, E., et al.: Astron. Astrophys. **520**, 61 (2010)
24. Charnley, S.B., et al.: Spectrochim. Acta **57**, 685 (2001)
25. Chiang, E.I., Goldreich, P.: Ap. J. **490**, 368 (1997)
26. Cho, J., Lazarian, A.: Ap. J. **669**, 1085 (2007)
27. Cortes, P.C., et al.: Ap. J. **628**, 780 (2005)
28. Crapsi, A., et al.: Astron. Astrophys. **470**, 221 (2007)
29. Crutcher, R.M., et al.: Ap. J. **514**, L121 (1999)
30. Daniel, F., et al.: MNRAS **363**, 1083 (2005)
31. Dartois, E., et al.: Astron. Astrophys. **399**, 773 (2003)
32. Demyk, K., et al.: Astron. Astrophys. **466**, 255 (2007)
33. Demyk, K., et al.: Astron. Astrophys. **489**, 589 (2008)
34. Di Francesco, J., et al.: Ap. J. **562**, 770 (2001)
35. Draine, B.T.: Ap. J. **636**, 1114 (2006)
36. Ducourant, C., et al.: Astron. Astrophys. **438**, 769 (2005)
37. Dupac, X., et al.: Astron. Astrophys. **404**, L11 (2003)
38. Dutrey, A., et al.: Astron. Astrophys. **286**, 149 (1994)
39. Dutrey, A., et al.: Astron. Astrophys. **309**, 493 (1996)
40. Dutrey, A., et al.: Astron. Astrophys. **325**, 758 (1997)
41. Dutrey, A., et al.: Astron. Astrophys. **402**, 1003 (2003)
42. Duvert, G., et al.: Astron. Astrophys. **355**, 165 (2000)
43. Evans II, N.J., et al.: Ap. J. **557**, 193 (2001)
44. Evans II, N.J., et al.: Ap. J. **626**, 919 (2005)
45. Falgarone, E., Puget, J.L.: Astron. Astrophys. **142**, 157 (1985)
46. Fuchs, G.W., et al.: Astron. Astrophys. **444**, 521 (2005)

47. Girart, J.M., Acord, J.M.P.: Ap. J. **552**, L63 (2001)
48. Girart, J.M., et al.: Ap. J. **525**, L109 (1999)
49. Girart, J.M., et al.: Science **313**, 812 (2006)
50. Goldreich, P., Kylafis, N.D.: Ap. J. **253**, 606 (1982)
51. Goldsmith, P.F., Langer, W.D.: Ap. J. **517**, 209 (1999)
52. Guélin, M., et al.: ApJS. **313**, 45 (2008)
53. Gueth, F., Guilloteau, S.: Astron. Astrophys. **343**, 571 (1999)
54. Gueth, F., et al.: Astron. Astrophys. **307**, 891 (1996)
55. Gueth, F., et al.: Astron. Astrophys. **323**, 943 (1997)
56. Gueth, F., et al.: Astron. Astrophys. **333**, 287 (1998)
57. Gueth, F., et al.: Astron. Astrophys. **375**, 1018 (2001)
58. Gueth, F., et al.: Astron. Astrophys. **401**, L5 (2003)
59. Guilloteau, S., Dutrey, A.: Astron. Astrophys. **291**, L23 (1994)
60. Guilloteau, S., Dutrey, A.: Astron. Astrophys. **339**, 467 (1998)
61. Guilloteau, S., et al.: Astron. Astrophys. **265**, L49 (1992)
62. Guilloteau, S., et al.: Astron. Astrophys. **478**, L31 (2008)
63. Guilloteau, S., et al.: Astron. Astrophys. **529**, 105 (2011)
64. Hamidouche, M., et al.: Ap. J. **651**, 321 (2006)
65. Harju, J., et al.: Astron. Astrophys. **482**, 535 (2008)
66. Harvey, D.W.A., et al.: Ap. J. **563**, 903 (2001)
67. Harvey, D.W.A., et al.: Ap. J. **596**, 383 (2003)
68. Hirano, N., Taniguchi, Y.: Ap. J. **550**, L219 (2001)
69. Hirano, N., et al.: Ap. J. **636**, L141 (2006)
70. Horne, K., Marsh, T.R.: MNRAS **218**, 761 (1986)
71. Hughes, A.M., et al.: Ap. J. **678**, 1119 (2008)
72. Hughes, A.M., et al.: Ap. J. **704**, 1204 (2009)
73. Isella, A., et al.: Astron. Astrophys. **469**, 213 (2007)
74. Isella, A., et al.: Ap. J. **701**, 260 (2009)
75. Jiménez-Serra, I., et al.: Ap. J. **603**, L49 (2004)
76. Jiménez-Serra, I., et al.: Ap. J. **650**, L135 (2006)
77. Kitamura, Y., et al.: Ap. J. **581**, 357 (2002)
78. Koerner, D.W., et al.: Icarus **106**, 2 (1993)
79. Kramer, C., et al.: Astron. Astrophys. **329**, L33 (1998)
80. Kruegel, E., Siebenmorgen, R.: Astron. Astrophys. **288**, 929 (1994)
81. Lada, C.J., et al.: Ap. J. **586**, 286 (2003)
82. Lai, S.-P., et al.: Ap. J. **598**, 392 (2003)
83. Lazarian, A.: J. Quant. Spectrosc. Radiat. Transf. **79**, 881 (2003)
84. Lee, C.-F., et al.: Ap. J. **576**, 294 (2002)
85. Lis, D.C., et al.: Ap. J. **571**, L55 (2002)
86. Lynden-Bell, D., Pringle, J.E.: MNRAS **168**, 603 (1974)
87. Mardones, D., et al.: Ap. J. **489**, 719 (1997)
88. Mathis, J.S., et al.: Astron. Astrophys. **217**, 425 (1977)
89. McCaughrean, M.J., et al.: Ap. J. **436**, L189 (1994)
90. Menten, K.M., et al.: Astron. Astrophys. **474**, 515 (2007)
91. Meny, C., et al.: Astron. Astrophys. **468**, 171 (2007)
92. Motte, F., André, P.: Astron. Astrophys. **365**, 440 (2001)
93. Motte, F., et al.: Astron. Astrophys. **336**, 150 (1998)
94. Natta, A., et al.: Astron. Astrophys. **416**, 179 (2004)
95. Ohashi, N., et al.: Ap. J. **475**, 211 (1997)
96. Ossenkopf, V., Henning, T.: Astron. Astrophys. **291**, 943 (1994)
97. Pagani, L., et al.: Astron. Astrophys. **467**, 179 (2007)
98. Palau, A., et al.: Ap. J. **636**, L137 (2006)
99. Pety, J., et al.: Astron. Astrophys. **458**, 841 (2006)
100. Piétu, V., et al.: Astron. Astrophys. **398**, 565 (2003)

101. Piétu, V., et al.: Astron. Astrophys. **443**, 945 (2005)
102. Piétu, V., et al.: Astron. Astrophys. **467**, 163 (2007)
103. Raga, A., Cabrit, S.: Astron. Astrophys. **278**, 267 (1993) (5 journal = Astron. Astrophys.)
104. Rodgers, S.D., Charnley, S.B.: Ap. J. **546**, 324 (2001)
105. Rodgers, S.D., Charnley, S.B.: Planet. Space Sci. **50**, 1125 (2002)
106. Roberts, H., Millar, T.J.: Astron. Astrophys. **364**, 780 (2000)
107. Roueff, E., et al.: Astron. Astrophys. **438**, 585
108. Salter, D.M., et al.: Astron. Astrophys. **492**, L21 (2008)
109. Schilke, P., et al.: Astron. Astrophys. **321**, 293 (1997)
110. Semenov, D., et al.: Ap. J. **673**, L195 (2008)
111. Shakura, N.I., Syunyaev, R.A.: Astron. Astrophys. **24**, 337 (1973)
112. Shirley, Y.L., et al.: Ap. J. **575**, 337 (2002)
113. Shu, F.H.: Ap. J. **214**, 488 (1977)
114. Shu, F.H., et al.: Ap. J. **370**, L31 (1991)
115. Simon, M., et al.: Ap. J. **545**, 1034 (2000)
116. Su, Y.-N., et al.: Ap. J. **671**, 571 (2007)
117. Tafalla, M., Bachiller, R.: Ap. J. **443**, L37 (1995)
118. Tafalla, M., et al.: Ap. J. **569**, 815 (2002)
119. Tamura, M., et al.: Ap. J. **525**, 832 (1999)
120. Tang, Y.-W., et al.: Ap. J. **700**, 251 (2009)
121. Terebey, S., et al.: Ap. J. **286**, 529 (1984)
122. Walawender, J., et al.: AJ. **132**, 467 (2006)
123. Walmsley, C.M., Ungerechts, H.: Astron. Astrophys. **122**, 164 (1983)
124. Wilkin, F.P.: Ap. J. **459**, L31 (1996)
125. Wyrowski, F., et al.: Ap. J. **514**, L43 (1999)
126. Zhou, S., et al.: Ap. J. **404**, 232 (1993)
127. Zhou, S., et al.: Ap. J. **466**, 296 (1996)

Index

© Springer-Verlag GmbH Germany, part of Springer Nature 2018
M. Dessauges-Zavadsky and D. Pfenniger (eds.), *Millimeter Astronomy*,
Saas-Fee Advanced Course 38, https://doi.org/10.1007/978-3-662-57546-8

Printed in the United States
By Bookmasters